Changing China: Migration, Communities and Governance in Cities

T0362188

China's unprecedented urbanization is underpinned by not only massive rural-urban migration but also a household registration system embedded in a territorial hierarchy that produces lingering urban-rural duality. The mid-1990s onwards witnessed increasing reliance on land revenues by municipal governments, causing repeated redrawing of city boundaries to incorporate surrounding countryside. The identification of real estate as a growth anchor further fueled urban expansion. Sprawling commodity housing estates proliferate on urban-rural fringes, juxtaposed with historical villages undergoing intense densification. The traditional urban core and work-unit compounds also undergo wholesale redevelopment. Alongside a large influx of migrants, major reshuffling of population has taken place inside metropolitan areas. Chinese cities today are more differentiated than ever, with new communities superimposing and superseding older ones. The rise of the urban middle class, in particular, has facilitated the formation of homeowners' associations, and poses major challenges to hitherto state dominated local governance.

The present volume tries to more deeply unravel and delineate the intertwining forms and processes outlined above from a variety of angles: circulatory, mobility and precariousness; urbanization, diversity and segregation; and community and local governance. Contributors include scholars of Chinese cities from mainland China, Hong Kong, Canada, Australia and the United States.

This volume was previously published as a special issue of *Eurasian Geography and Economics*.

Si-ming Li is currently Director of the David C. Lam Institute of East-West Studies and Chair Professor of Geography, Hong Kong Baptist University.

Shenjing He is an Associate Professor at the Department of Urban Planning and Design, The University of Hong Kong.

Kam Wing Chan is Professor of Geography at the University of Washington.

Changing China: Migration, Communities and Governance in Cities

Edited by
Si-ming Li, Shenjing He and Kam Wing Chan

LONDON AND NEW YORK

First published 2017
by Routledge
2 Park Square, Milton Park, Abingdon, Oxon, OX14 4RN, UK

and by Routledge
711 Third Avenue, New York, NY 10017, USA

First issued in paperback 2018

Routledge is an imprint of the Taylor & Francis Group, an informa business

British Library Cataloguing in Publication Data
A catalogue record for this book is available from the British Library

ISBN 13: 978-1-138-33236-2 (pbk)
ISBN 13: 978-1-138-69084-4 (hbk)

Typeset in Times New Roman
by RefineCatch Limited, Bungay, Suffolk

Publisher's Note
The publisher accepts responsibility for any inconsistencies that may have
arisen during the conversion of this book from journal articles to book chapters,
namely the possible inclusion of journal terminology.

Disclaimer
Every effort has been made to contact copyright holders for their permission to
reprint material in this book. The publishers would be grateful to hear from any
copyright holder who is not here acknowledged and will undertake to rectify
any errors or omissions in future editions of this book.

Contents

CONTENTS

Citation Information

The chapters in this book were originally published in *Eurasian Geography and Economics*, volume 55, issue 4 (August 2014) and volume 56, issue 3 (June 2015). When citing this material, please use the original page numbering for each article, as follows:

Chapter 2
Instability of migrant labor supply in China: evidence from source areas for 1987–2008
Yan Yuan, Zhao Rong, Rudai Yang and Liu Yang
Eurasian Geography and Economics, volume 56, issue 3 (June 2015) pp. 231–259

Chapter 3
Residential mobility within Guangzhou city, China, 1990–2010: local residents versus migrants
Si-ming Li and Yushu Zhu
Eurasian Geography and Economics, volume 55, issue 4 (August 2014) pp. 313–332

Chapter 4
Participation and expenditure of migrants in the illegal lottery in China's Pearl River Delta
Zhiming Cheng, Russell Smyth and Gong Sun
Eurasian Geography and Economics, volume 55, issue 4 (August 2014) pp. 333–361

Chapter 5
The effects of residential patterns and Chengzhongcun housing on segregation in Shenzhen
Pu Hao
Eurasian Geography and Economics, volume 56, issue 3 (June 2015) pp. 308–330

Chapter 6
Space to maneuver: collective strategies of indigenous villagers in the urbanizing region of northwestern China
Jing Song
Eurasian Geography and Economics, volume 55, issue 4 (August 2014) pp. 362–380

Chapter 7

Neighborhood conflicts in urban China: from consciousness of property rights to contentious actions
Qiang Fu
Eurasian Geography and Economics, volume 56, issue 3 (June 2015) pp. 285–307

Chapter 8

Homeowner associations and neighborhood governance in Guangzhou, China
Shenjing He
Eurasian Geography and Economics, volume 56, issue 3 (June 2015) pp. 260–284

Chapter 9

Creating and defending concept of home in suburban Guangzhou
Dan Feng, Werner Breitung and Hong Zhu
Eurasian Geography and Economics, volume 55, issue 4 (August 2014) pp. 381–403

For any permission-related enquiries please visit:
http://www.tandfonline.com/page/help/permissions

Migration, communities and governance in Chinese cities: unfolding new forms and processes

Si-ming Li[a], Kam Wing Chan[b] and Shenjing He[c]

[a]Hong Kong Baptist University; [b]University of Washington; [c]University of Hong Kong

Epochal migration and mobility

China's unprecedented urbanization is unique in world development history, not only because of its rapidity and the sheer scale of rural-urban migration, but also because of its special institutions of handling rural and urban populations. The *hukou* system, in particular, delineates every individual in the country according to means of subsistence – agricultural (rural) versus non-agricultural (primarily urban) – and place of official residence, and helps erect highly impermeable "invisible walls" not only between town and country but also within major cities (Chan 1994; 2010). Starting from a base of below 20 per cent when the reform began in the late 1970s, China's level of urbanization surpassed the 50 per cent mark in 2011 for the first time in history, the latest figure being 54.8 per cent in 2014 (State Council 2015). However, this rate remains well below those of the United States and other economically advanced countries; moreover, "urbanization" in China is also far from complete in that out of the total urban population of 750 million in 2014, about 250 million do not hold the *hukou* in their current place of domicile (State Council 2015). In fact, the percentage of non-*hukou* residents in urban places has been increasing incessantly – and alarmingly – throughout the reform period, from a couple of per cents to about one-third of the urban population nowadays (Chan 2014).

These non-*hukou* urban residents are denied access to a wide range of public or social services, such as schooling, healthcare and social housing. The denial of local *hukou* status also traps rural migrants to the lower end of the urban labour market. The great majority of them can only find work in construction sites and factories performing original equipment manufacturing requiring little skills, or in low-end service jobs in restaurants, massage parlours, and warehouses and other transport activities. Quite a few have to make their living through scavenging. Clearly, all these jobs lack even the basic security. In short, the lack of citizenship rights in major urban areas renders rural migrants in a state of disenfranchisement, marginalization, precariousness and predicaments (Wong et al. 2007). Irrespective of the huge difference in income levels and prospects for advancement between the glittering host metropolis and home village, studies have revealed that only a relatively few migrants indicate an intention to establish permanent residence in the host city (Du and Li 2012). The lack of interest could well be an outcome of the immense legal and other obstacles they face in the hostile urban environment. Circular and repeated migration is common, if not the norm; so are "split families", with often wives and/or children being left behind in the home village. These arrangements are far from ideal, and often at high costs to the family and children's education (Chan 2015).

1

The deeply ingrained urban-rural duality is further underpinned by a territorial administrative hierarchy that channels financial and other resources to leading metropolises, and contributes to landed properties based urban expansion that has fundamentally transformed former compact cities to sprawling metropolitan areas spanning hundreds of kilometres across (Lin 2007). Severe and often violent contestations surrounding land acquisitions on ever-receding urban-rural fringes feature prominently on news media (Hsing 2010). Devolution of government decision-making power under marketization has given rise to a rather peculiar local state "developmentalism" (Oi 1995; Zhu 1999; Pei 2006). Yet, the 1994 tax-sharing reform has re-concentrated fiscal resources at the centre in Beijing. Land leasing and related revenues have become the single-most important financial means for municipal governments to achieve their economic goals. Instead of providing a more-levelled playing field, the heavy use of market tools, including the commodification of urban land and housing and the gradual dismantling of work-unit compounds as residential communities, has aggravated socio-spatial inequalities within and beyond major cities (He and Wu 2009).

To date, reforms on the *hukou* system and the associated institutions of land tenure and social welfare provisions have mostly benefited the rich and the highly educated such as successful business migrants and selected university graduates. These new measures have further differentiated the urban population socially, politically and spatially, especially in Beijing, Shanghai and other so-called "first-tier" cities (Li et al. 2012). The recent call for conferring 100 million new urban *hukou* by 2020, which is spelled out in China's *New-type Urbanization Plan* and which also stipulates a target urbanization rate of 60% by the same year and improved housing conditions for low-income groups and education for migrant children, is unlikely to fundamentally change the nature of socio-spatial differentiation in the leading metropolises where most migrants congregate, however. This is because the new policy applies primarily to small urban places. Indeed, there are still bugging worries that the new urbanization plan might simply be hijacked by local governments for their own agenda (Chan 2014).

Earlier studies on migration in China tended to focus on the spatial patterns and changing process of rural-urban migration as well as migrants' marginalization status (Chan and Zhang 1999; Fan 2007; Shen 1999; Wang and Fan 2006; Wu 2008). Recent research has extended to analyzing migrants' social integration, or the lack thereof, with the urban society (Wang and Fan 2012), their social networks (Liu et al. 2012; Yue et al. 2013), access to urban homeownership (Wu and Wang 2014); residential satisfaction (Tao et al. 2014), and community attachment and sense of belonging to the city (Wu 2012; Du and Li 2012). The scope has continued to broaden, covering topics such as migrant children (Fong 2006) and families (Fan et al. 2011), and inter-generational differentiation among rural migrants (Yue et al. 2010). As the number of rural migrants and their contribution to China's urban "miracle", or some would argue "disaster" rather, continues to increase, this social group has received more media and scholarly attention.

Heterogeneity in socioeconomic composition and migrant housing

Much has been written on "villages-in-city" or for some authors, "urban villages", another major manifestation of China's rural-urban duality (Chung 2014; Wang et al. 2014). Villages on city outskirts are rural entities and fall outside the purview of the urban governance system. Instead, they are administered by village committees and the supervising township governments, the two levels of rural administration constituting what are called rural collectives. As direct descents of the former People's Communes, these collectives assume both economic and administrative functions, including ownership of land within their jurisdiction. By law, rural land cannot be transacted for urban uses. Only the municipal government has the right to requisition rural land and turn it to state-owned land for leasing under the system of paid transfer of land use rights (Xu et al. 2009). The lure of windfall leasing revenues drives municipal governments to engage in frenetic "land enclosure" exercises, resulting in rapid engulfing of suburban villages by urban development (hence the notion of villages-in-city) and serious local debt crisis (Tsui 2011).

Many villagers have built new houses on nearby land plots for their own use, and the vacated old village houses are often rented to migrants from afar but working in nearby towns or cities. Some villagers have also added floor space to generate further rental incomes, often with little regard to building safety and ventilation because construction in rural villages is not subject to municipal building and public health codes. In Guangzhou and Shenzhen, residential structures without elevators rising to nine storeys or higher erected on both sides of narrow alleys are not uncommon in "villages-in-city". Ventilation is poor; sunlight can hardly penetrate the narrow alleys; and fire hazards are high, as fire engines cannot make their way through the alleys. Most renters in these villages are migrant labourers from rural areas afar who are neither eligible for work-unit and municipal social housing nor able to afford private (commodity) housing. Such villages have effectively become migrant enclaves, with outsiders easily out-numbering locals by a large ratio. Despite intermingling in confine space, migrant-renters and villager-landlords seldom interact. Instead, mutual mistrust prevails between the two groups (Du and Li 2010). The hastily constructed structures often lack proper maintenance and deteriorate rapidly. Many villages-in-city have disintegrated into becoming "slums", marked by high crime rates and grossly inadequate infrastructure services.

As economic-cum-political entities, many rural collectives in suburban areas have been transformed to share-holding companies. While industrial township and village enterprises were the major non-agricultural undertakings in suburban villages in the 1980s and early 1990s, the main concerns of township and village authorities these days are landed-property developments (Hsing 2010). Bargaining for better land requisition compensations and distributing the proceeds to individual villagers fairly is a major responsibility of the village authority. But township and village authorities would try to generate land-based incomes more directly. For example, in Beijing and other cities, suburban township governments have introduced a type of commodity housing with *xiao-chanquan* ("limited property right"),[1] which is sold at prices substantially below those of housing built on properly leased land. The questionable legality of such housing notwithstanding, the number of units in this category, according to unofficial estimates, astoundingly reaches 70 million (Ren Zhiqiang 2013). Buyers of *xiao-changquan* housing are often migrants who cannot afford to buy in the "proper" commodity housing sector (Hsing 2010).

Aside from living in villages-in-city and *xiao-changquan* housing in outlying suburbs, migrants also congregate in construction sites, factory dormitories, dilapidated private housing predating the Communist takeover in 1949, and illegal basements and former underground air-raid bunkers (Wu 2004; Johnson 2015). Some have infiltrated into former work-unit compounds by renting privatized work-unit housing whose owners have moved to better city condominiums or suburban apartments. Irrespective of the enlarged spatial choice set, migrants, especially low-status migrant workers from rural areas, still face immense difficulties in establishing themselves in the city. Even migrants who are doing well and who have the money are barred from purchasing homes in the private market in many large cities as local governments attempt to cool down the overheated housing market in recent years (*Global Times* 2013).

China's more than thirty years of double-digit economic growth has been accompanied by an alarmingly rapid rise in income and wealth inequalities. A small number of migrants have succeeded in moving up the socio-economic ladder. Large numbers of arrivals to Beijing, Shanghai and other metropolises are graduates from leading universities with degrees in professional fields. The popularization of the term *yizu* or "ant tribe", referring to those young migrant graduates in major cities who have to live like peasant-migrants in subdivided inner-city tenements and substandard housing in villages-in-city, pinpoints to the increasing difficulty for formal *hukou* attainment even for migrants with a university degree (Li et al. 2012). Nonetheless, over the years quite a few migrant YUPPIES (young upwardly-mobile professionals), especially those working in state and quasi-state sectors such as universities and research institutions, have managed to get hold of the proper *hukou*. The futures seem bright for these migrant YUPPIES; however, they are still subject to constraints quite different from their local counterparts. As a case in point, large percentages of home purchases in both Guangzhou and Shanghai were made with parental support (Li 2010). However, evidently for young professional migrants from lagging provinces and rural areas, parental support is mostly out of the question. In Guangzhou, there is evidence that buyers of more reasonably priced apartments in far-off suburbs are mainly migrant YUPPIES who communicate in Mandarin Chinese rather than Cantonese, the local dialect. Literacy in information technology enables them to form online-offline communities to pursue communal goals (Li and Li 2014).

Heterogeneity and socio-spatial differentiation

Increased residential mobility and widening socio-economic inequality have resulted in spatial differentiation and residential segregation (Huang and Li 2014; Li 2012). The rise of gated communities and enclave urbanism is a vivid manifestation of spatial segregation and fragmentation (He 2013; Pow 2009). In Chinese cities, socio-spatial inequality and residential segregation have been commonly observed and measured at different scales, especially at the neighbourhood level (see Li and Wu 2008; He et al. 2010). Unlike the experience of North American and some European cities, issues of race and ethnicity rarely play a role in the dynamics of socio-spatial differentiation in urban China except in the "autonomous regions" in the northwest and southwest frontiers. Instead, it is the rural-urban divide and income and wealth disparity that work to produce residential segregation. Large-scale redevelopment generates another dynamic to intensify residential segregation through constant reworking of the urban fabrics (He 2013; He and Wu 2005). The distribution of various social groups has been reshuffled, and the outcome of stark social differentiation is manifested at different spatial scales. For instance, Beijing's socio-spatial structure features clear socio-spatial differentiation between residential neighbourhoods as well as the mixing of diverse social groups at the "city district" level[2] (Feng et al. 2008). In general, the process of socio-spatial restructuring in Chinese cities typically involves rapidly gentrifying inner cities and sprawling suburbs with middle-class homes, industrial zones and rural migrant settlements juxtaposed to each other. Resultantly, layering fabrics from different historical periods have woven out a patchwork of various types of enclaves, comprising work-unit compounds, gated communities and low-income villages-in-city (He 2013; Qian 2014).

Villages-in-city and upscale gated communities represent two extreme spatial manifestations of increasing polarization in Chinese cities. Above we discussed in some detail the former. Regarding the latter, in China they are often deemed as an ideal way of living, with high-quality built environment, tight security measures, and well-organized services and management (Wu 2004; Xu and Yang 2009). In contrast, villages-in-city are invariably chaotic and crime-prone (Hao et al. 2013). The social meanings attached to these two forms of enclaves also greatly differ. Enclave living in gated community housing estates is an expression of a safe, "high-quality", and privileged lifestyle, while villages-in-city are makeshift shelters to eke out an urban living for those who have little choice (He 2013; Liu et al. 2010). Nonetheless, it is also not uncommon that constant interactions take place between these two contrasting kinds of enclaves. This is because convenient and low-cost services and goods provided by villages-in-city often meet the needs of neighbouring gated communities (Breitung 2012; Li et al. 2012). Hence, the wholesale redevelopment of these villages not only destroys the shelter and livelihood of rural migrants, but also dissolves the social mix particularly in the expanding suburbs, thus exacerbating social differentiation and residential segregation. In this regard, we advocate a closer examination of socio-spatial differentiation at multiple scales, especially at a finer scale, taking into account also various social, cultural and historical considerations.

Communities and local governance

As a microcosm of the Chinese urban society at the local level, community becomes an increasingly important social and spatial unit to understand the latest urban transformation. The state is still paramount in China through its social engineering, economic and urban planning, and social control, despite the ongoing market transition (see Bian and Logan 1996; Chan 2009; He and Wu 2009). However, societal forces in China's urban transformation should not be overlooked either. Since the 1990s, with the dissolution of the cellular spatial structure under the work-unit or *danwei* system, *xiaoqu* (residential neighbourhood), mainy gated, has emerged as a potential site for creating a new quasi-public sphere. Meanwhile, against the backdrop of frenetic urbanization, the rising consciousness of property rights (Davis and Lu 2003; Read 2003), the emergence of consumer citizens (Davis 2006) and the increased personal freedom and mobility (Li 2003) have contributed to an increasingly diverse urban society and instigated various forms of contestations in urban areas. Property rights have been recognized as an important lens to comprehend conflicts and unrests accompanying the rapid urbanization. Within the recent decade, rights defending (*weiquan*) activism and protests have seen a heightened visibility in China studies (e.g. Cai 2010; Lee 2008; O'Brien and Stern, 2008).

At the neighbourhood level, restored private property rights and the rising demand of homeowners for taking part in local decision-making have bred activism and different forms of grassroots response to defend property against the misconducts of some powerful players (Davis 2006; Read 2003, 2008; Tomba 2005; Yip and Jiang 2011). Meanwhile, the prevalence of newly-built gated communities and the enactment of property rights law in 1997 have given rise to some civic organizations at the neighbourhood level, including property management companies and homeowners' associations. However, the state continues to penetrate local communities through its urban and cultural policies to educate, civilize, and rectify people, e.g. the "community construction" campaign. The campaign is an attempt to strengthen the role of street offices and residents' committees in steering and overseeing social and cultural exchanges at the grassroots level (Bray 2006). The formation and continual functioning of homeowners' associations also depends largely on the support of the municipal government, and is under tight scrutiny of the relevant residents' committee and street office (Tomba 2008; Fu and Lin 2014). As such, these newly emerging social organizations have only a limited role in fostering genuine civil resistance in China (Tomba 2009; Fu and Lin 2014). The increasing prevalence of gated communities can thus be seen as a result of the downscaling of urban governance and the enforcement of political control (Huang 2006). Such considerations notwithstanding, the perspective from local communities and ground-up research approaches do have potential to add fresh insights to the critical assessment of current Chinese urbanism.

Highlights of the following chapters

The intertwining processes of migration, mobility and community change in Chinese cities outlined above are further illustrated in the following chapters.

Using the 1986–2008 annual household survey data provided by the Research Center on Rural Economy under China's Ministry of Agriculture, Yuan et al. (2015) delve into a comparatively under-researched issue, i.e. the circular nature and the instability problem of China's rural-urban migration. Owing to various institutional restrictions, circular rural-urban migration has become prevalent and "permanent", resulting in unstable migrant labour supply. Making use of the longitudinal data, the paper investigates how the instability evolved over time and varied across regions. Despite some slight improvement, the instability persists because the discriminatory *hukou* system remains largely unchanged. Their findings suggest that variables such as educational attainment, labour surplus, and network effects are correlated with the onset of migration and also migration (in)stability. In addition, migration instability is also attributed to local migration regulations and individual characteristics. For instance, young healthy males are more likely to engage in repeated migration than other types of labourers. The study adds to our understanding of the dynamics of the parallel existence of rural labour surplus and labour shortages, which was analyzed earlier by Chan (2010). It also updates our understanding of several important issues, including migrant labourers' human capital accumulation and remuneration, the impact of the Labour Contract Law on migration stability, regional differences in migration propensities, and the reasons for relocating firms from coastal to inland areas.

Li and Zhu take on board the issue of migrants' housing by comparing the propensity to move residences as well as the factors underlying these residential moves between migrants and locals in Guangzhou. The findings show that among those with local *hukou*, mobility rates declined in the 2000s after an initial period of relatively high mobility, which could be attributed to the high rate of homeownership resulting from discounted sale of work-unit housing as well as rapidly rising home prices. The locals have now settled into a more stable community, as one would expect almost two decades after the housing reform. However, for migrants without the Guangzhou *hukou*, the mobility rate has continued to stay high. Non-*hukou* residents of major cities are indeed mobile not only moving back and forth between the home village and the city of current domicile but moving from one residence to another within the same city. In many instances the move takes place in the same sphere of informal housing mainly in villages-in-city, or from a more centrally located village to a much more remote one due to redevelopment. Such continuing "mobility" actually reflects migrants' predicament of not being able to settle where they work under China's current institutions. Instead of moving up with increased geographic mobility as has been hypothesized in the conventional economic theory, these movements often tell of a downward spiral to poverty of many migrants, as they face greater and greater obstacles in the city to find affordable living quarters that allow them to carry on their life, i.e. reasonable access to jobs and to schools (for their kids).

Along the broad downward trajectory, many migrants are trapped in another vicious cycle. Loneliness, lack of social support, being free from surveillance by parents, spouse and older-generation kin in the home village, and persuasion from peers are all contributing factors to migrants' propensity to gamble and participate in other socially unsanctioned, if not illegal, behaviours that are likely to keep them at the bottom of the society. The paper by Cheng et al. (2015) examines migrants' participation in an illegal lottery in the Pearl River Delta based on a survey conducted in 2006. The findings show that being male and living in factory dormitory significantly heightens the propensity to participate in illegal lottery. On the one hand, those migrants who are unwilling to give up their farmland in the home village are more reluctant to take part in the game. To address the issue of persistent rural-urban dichotomy, many economists have advocated doing away with the difference between state and collective land ownership to allow unrestricted circulation of rural land. While an integrated rural-urban land market should be the goal of a true market economy, which China aims to establish, rapid marketization of rural housing land runs the risk of mass dispossession and dislocation of the peasantry especially in regions where the protection of property rights is weak (Chan 2014). The Cheng et al. findings can be seen as words of caution on such a neoliberal policy prescription. Migrants' precarious situation to which the chapter devotes also prompts us to think about the vicious cycle many migrants of the opposite sex have unfortunately fallen into: prostitution. While the seriousness of this problem is obvious to even a causal visitor to the Pearl River Delta, systematic research remains wanting.

Hao addresses another pressing issue of residential segregation in Shenzhen from two angles: segregation between the privileged *hukou* holders and underprivileged non-*hukou* migrants as well as the spatial separation of formal urban housing from *chengzhongcun* ("villages-in-city"). Drawing on two sets of detailed demographic and building data from the Household Registration Database and Municipal Building Database, his study measures residential segregation at different geographical and administrative scales. Three patterns of residential segregation are evident: the population separation between the SEZ (Special Economic Zone) and the non-SEZ districts caused by the different development paths, the segregation among sub-districts resulting from uneven economic development and associated employment opportunities, and the separation at the level of residents' committees owing to the availability of specific housing types. On a different dimension, non-*hukou* migrants are largely segregated from *hukou* holders due to their much-constrained choice of housing and the widespread availability of *chengzhongcun*. In contrast, the degree of segregation at the sub-district level is relatively low, which maintains a spatially more equitable setting that enables migrants to reside within short distances to access jobs and amenities. The study warns that ongoing large-scale urban renewal programs targeting *chengzhongcun* are most likely to jeopardize the existing social mix at the sub-district level, and may eventually aggravate segregation at larger geographical scales.

Suburban rural collectives in major coastal cities may be the first to introduce measures such as the issuance of *xiao-chanquan* to derive greater economic benefits from rising land values. But such grassroots initiatives, which could be traced back to experiments on contracting land plots to individual peasant-households in certain localities in Anhui Province in the late 1970s, have been distinct features of the Chinese reform from the outset. Oi (1992) coins the term "local state corporatism" to characterize the entrepreneurial predispositions of grassroots authorities in rural China in early reform times. Evidently, location specificity is an inherent feature of local state corporatism. The Go-West Policy launched in 2000 to address worsening regional inequalities has to some extent enhanced development potentials in the Western Frontier Region, where large populations of ethnic minorities are found. Arguing that urbanization in China is sensitive to and dependent on local conditions, Song (2015) in this issue analyses how the authorities and peasants of two villages on the outskirts of Yinchuan, capital of Ningxia Hui Autonomous Region, devise strategies that tend to vary with the distance to the city centre so as to ride the tide of urbanization. Agency is essential for indigenous villagers not to end up in a state of deprivation and hopelessness in the face of urban sprawl. It also shows that some locals and grassroots have figured out ways to protect their interests despite the rigid institutional structure.

Both the Fu and He's chapters look at the emerging issue of neighbourhood governance based on a large-scale household survey in Guangzhou, China. These two chapters have different emphases, however.

In the existing literature, while the effects of neighbourhood conflicts on contentious actions in urban China have been well documented, the causal link between neighbourhood perception and neighbourhood conflicts has not been fully explored. Drawing on data from the said household survey, Fu uses interview materials to examine the structure, determinants and consequences of neighbourhood conflicts. His study opens a new dimension in understanding neighbourhood conflicts by revealing the close link between neighbourhood perception and neighbourhood conflicts. On the one hand, consciousness of property rights has significant effects on the perception of neighbourhood conflicts; for example, residents with higher levels of consciousness of property rights may perceive more neighbourhood conflicts and eventually advocate neighbourhood activism. On the other hand, neighbourhood conflicts can reinforce residents' consciousness of property rights. In other words, neighbourhood perception shaped by consciousness of property rights and the effect of neighbourhood conflicts on consciousness of property rights are two dynamics, which are mutually reinforcing one another, that determine contentious actions in urban China. Neighbourhood conflicts with local and grassroots governments are most likely to be translated into contentious actions, i.e. contentious actions are prone to occur when local governments cannot provide economic prosperity or social security, or fail to adhere to rules stipulated by the central authority. Data in this study suggest that contentious politics in China are to a great extent shaped by the boundary of institutional protocols and state regulations, which largely conforms to the rule consciousness framework (Perry 2008).

Complementing Fu's work, He's chapter examines the formation and operation of homeowner association (HOA) and its governance efficacy in urban neighbourhoods through the analytical lenses of private governance and collective action. Her study also links the empirical analyses with the broader state-market-society interactions in China. Her chapter makes three points. First, there is a persisting state influence throughout the development of HOAs. To a large extent, the establishment and performance of a HOA is contingent upon the legislative system, the disputes and power relation between the HOA and state agents, as well as the involvement of state agents in neighbourhood activities. Second, although the HOA, formed by homeowners, is physically an integrated part of the gated community (private neighbourhood), it is far from any form of private governance by Western definition. Interestingly, it has become a societal force to counterbalance the market force brought by the property management company and the state power enforced by the residents' committee, to address homeowners' political and material needs. Third, the effectiveness of the HOAs in neighbourhood governance can be largely explained by the collective action theory, which mainly concerns endogenous factors. Yet, in the Chinese context exogenous factors such as state power and market force also bring a strong imprint on the decision-making of collective action in terms of determining organizational arrangements, the availability and quality of information, and the benefits of collective action.

Turning to the brighter side of China's recent changes, the chapter by Feng et al. examines lives in Lijiang Garden, a sprawling housing estate comprising high-rise apartment blocks located some 10 km to the south of Zhujiang New Town, the new central business district of Guangzhou, and Shunde Country Garden, a huge residential development comprising single-family homes located more than 50 km to the southwest of central Guangzhou. These are homes for the middle class and upper middle class. It is useful to note that Lijiang Garden is one of the communities with active online-offline forums studied by Li and Li (2014). Whereas migrant YUPPIES dominate the population of Lijiang Garden, a high percentage of Shunde Country Garden residents are retirees from Hong Kong. With much improved road and rail networks in recent years, long-distance "commuting" to Guangzhou is now much more viable. Thus, Shunde Country Garden increasingly assumes the character of a dormitory community of Hong Kong. Notwithstanding the difference in location and population composition, in both Lijiang Garden and Shunde Country Garden, the chapter discerns distinct physical and mental boundaries that separate residents of the estate from residents of surrounding villages, the latter consisting of both indigenous villagers and migrants from afar. Efforts by the developer and estate management firm to construct images of an ideal home have helped to sustain the rather uneasy juxtaposition of population groups who live in close proximity. Nevertheless, the notion of "home" and the related concept of family togetherness deserve further investigation to tease out more of their implications for people, life and community, given China's continuing *hukou*-based segregation and rising urban housing prices.

The chapters included in the volume collectively present studies of migration and its impacts, and urban communities, especially in relation to socio-spatial differentiation, diversity and governance. We hope to continue to push the frontiers of research on mobility and community in China, leading to more complete and accurate understanding of the prodigious rural-urban migration, a process that not only has changed China, but also is changing the world.

Acknowledgements

All chapters in this volume are from papers first published in *Eurasian Geography and Economics* (*EGE*) in two special issues (Volume 55, No. 4, 2014, and Volume 56, No. 3, 2015), respectively. The introductory chapter is revised from the two introductory papers of the special issues. We would like to thank Peter Gane who on behalf of Routledge, the publisher of the journal, invited us to publish the special issues in the form of a book, and all contributors for allowing us to republish their work in the present volume. Reviews and editing were carried out by the *EGE* editors and invited referees; many of the papers went through two to three revisions by authors before acceptance. Thanks are particularly due to Nancy Place, who meticulously copyedited all the manuscripts with professionalism. Out of the eight chapters to follow, six were presented at a workshop held at Hong Kong Baptist University in November 2013, in which all of us participated. We would also like to acknowledge the university's financial support.

Note

1. Also known as *"xiang chanquan,"* or housing with township-issued ownership certificates.
2. Many city districts include large sketches of rural areas and rural population and because of that, those areas are often much larger than the "metropolitan areas" as they are commonly used in the West (see Wang and Chan 2014).

References

Bai, Xuemei, Shi Peijun, and Liu Yansui. 2014. "Society: Realizing China's Urban Dream." *Nature* 509(7499): 158–160.

Bian, Yanjie, and John R. Logan. 1996. "Market Transition and the Persistence of Power: The Changing Stratification System in Urban China." *American Sociological Review* 61(5): 739–758.

Bray, David. 2006. "Building 'Community': New Strategies of Governance in Urban China." *Economy and Society* 35(4): 530–549.

Breitung, Werner. 2012. "Enclave Urbanism in China: Attitudes Towards Gated Communities in Guangzhou." *Urban Geography* 33(2): 278–294.

Cai, Yongshun. 2010. *Collective Resistance in China: Why Popular Protests Succeed or Fail.* Stanford, Calif.: Stanford University Press.

Chan, Kam Wing. 1994. *Cities with Invisible Walls: Reinterpreting Urbanization in Post-1949 China.* New York: Oxford University Press.

Chan, Kam Wing. 2009. "The Chinese Hukou System at 50." *Eurasian Geography and Economics* 50(2): 197–221.

Chan, Kam Wing. 2010. "A China Paradox: Migrant Labor Shortage amidst Rural Labor Supply Abundance." *Eurasian Geography and Economics* 51(4): 513–530.

Chan, Kam Wing, 2014. "China's Urbanization 2020: A New Blueprint and Direction." *Eurasian Geography and Economics* 55(1): 1–9.

Chan, Kam Wing, 2015. "China's Missing Children." *South China Morning Post*, December 16, A15.

Chan, Kam Wing, and Zhang Li. 1999. "The Hukou System and Rural-Urban Migration in China: Processes and Changes." *The China Quarterly* 160: 818–855.

Chung, Him. 2014. "Rural Transformation and the Persistence of Rurality in China." *Eurasian Geography and Economics* 54: 594–610.

Davis, Deborah S. 2006. "Urban Chinese Homeowners as Consumer-Citizens." In *The Ambivalent Consumer*, edited by S. Garon and P. Maclachlan, 281–299. Ithaca, NY: Cornell University Press.

Davis, Deborah S., and Hanlong Lu. 2003. "Property in Transition: Conflicts over Ownership in Post-Socialist Shanghai." *European Journal of Sociology* 44(1): 77–99.

Du, Humin, and Si-ming Li. 2010. "Migrants, Urban Villages, and Community Sentiments: A Case of Guangzhou, China." *Asian Geographer* 27: 93–108.

Du, Huimin, and Si-ming Li. 2012. "Is it Really Just a Rational Choice? The Contribution of Emotional Attachment to Temporary Migrants' Intention to Stay in the Host City in Guangzhou." *China Review* 12(1): 73–93.

Fan, Cindy C. 2007. *China on the Move: Migration, the State, and the Household*. Abingdon, Oxon: Routledge.

Fan, Cindy C., Mingjie Sun, and Siqi Zheng. 2011. "Migration and Split Households: A Comparison of Sole, Couple, and Family Migrants in Beijing, China." *Environment & Planning A* 43(9): 2164–2185.

Feng, Jian, Fulong Wu, and John Logan. 2008. "From Homogenous to Heterogeneous: The Transformation of Beijing's Socio-Spatial Structure." *Built Environment* 34(4): 482–498.

Fong, Vanessa L. 2006. Chinese youth between the margins of China and the First World. In: *Chinese Citizenship: Views from the Margins*, edited by Vanessa L. Fong and Rachel Murphy, 141–173. London: Sage.

Fu, Qiang, and Nan Lin. 2014. "The Weaknesses of Civic Territorial Organizations: Civic Engagement and Homeowners Associations in Urban China." *International Journal of Urban and Regional Research* 38(6): 2309–2327.

Global Times. 2013. "Stricter Policy to Cool Down Real Estate Market." March 3. http://www.globaltimes.cn/content/765401.shtml. Accessed January 22, 2015.

Guojia Xinxing Chengzhenhua Guihua (2014–2020) (National New-type Urbanization Plan). 2014. March 17. Accessed March 19, 2014. http://politics.people.com.cn/n/2014/0317/c100124649809.html

Hao, Pu, Pieter Hooimeijer, Richard Sliuzas, and Stan Geertman. 2013. "What Drives the Spatial Development of Urban Villages in China?" *Urban Studies* 50(16): 3394–3411.

He, Shenjing, and Fulong Wu. 2005. "Property-led Redevelopment in Post-reform China: A Case Study of Xintiandi Redevelopment Project in Shanghai." *Journal of Urban Affairs* 27(1): 1–23.

He, Shenjing, and Fulong Wu. 2009. "China's Neoliberal Urbanism: Perspectives from Urban Redevelopment." *Antipode* 41: 282–304.

He, Shenjing, and Fulong Wu. 2009. "China's Emerging Neoliberal Urbanism: Perspectives from Urban Redevelopment." *Antipode* 41(2): 282–304.

He, Shenjing, Yuting Liu, Fulong Wu, and Chris Webster. 2010. "Social Groups and Housing Differentiation in China's Urban Villages: An Institutional Interpretation." *Housing Studies* 25(5): 671–691.

He, Shenjing. 2013. "Evolving Enclave Urbanism in China and its Socio-spatial Implications: The Case of Guangzhou." *Social & Cultural Geography* 14(3): 243–275.

Hsing, You-tien. 2010. *The Great Urban Transformation: Politics of Land and Property in China.* Oxford: Oxford University Press.

Huang, Youqin, and Si-ming Li (Eds.). 2014. *Housing Inequality in Chinese Cities.* Abingdon, Oxon: Routledge.

Huang, Youqin. 2006. "Collectivism, Political Control, and Gating in Chinese Cities." *Urban Geography* 27(6): 507–525.

Johnson, Ian. 2015. "The Rat Tribe of Beijing." *Al Jazeera America*, January 24. Accessed February 4. http://projects.aljazeera.com/2015/01/underground-beijing/.

Lee, Ching Kwan. 2008. "Rights Activism in China." *Contexts* 7(3): 14–19.

Li, Limei, Si-ming Li, and Yingfang Chen. 2012. "Better City, Better Life, But for Whom? The Hukou and Resident Card System and the Consequent Citizen Stratification in Shanghai." *City, Culture and Society* 1: 145–154.

Li, Limei, and Si-ming Li. 2014. "Living the Networked Life in the Commodity Housing Estates: Everyday Use of Online Neighborhood Forums and Community Participation in Urban China." In: *Housing Inequalities in Chinese Cities*, edited by Youqung Huang and Si-ming Li, 181–198. London and New York: Routledge.

Li, Si-Ming. 2003. "Housing Tenure and Residential Mobility in Urban China: A Study of Commodity Housing Development in Beijing and Guangzhou." *Urban Affairs Review* 38(4): 510–534.

Li, Si-ming. 2010. "Mortgage Loan as a Means of Home Finance in Urban China: A Comparative Study of Guangzhou and Shanghai." *Housing Studies* 25: 857–876.

Li, Si-Ming. 2012. "Housing Inequalities under Market Deepening: The Case of Guangzhou, China." 44(12): 2852–2866.

Li, Si-ming, Yushu Zhu, and Limei Li. 2012. "Neighborhood Type, Gatedness, and Residential Experiences in Chinese Cities: A Study of Guangzhou." *Urban Geography* 33(2): 237–255.

Li, Zhigang, and Fulong Wu. 2008. "Tenure-based Residential Segregation in Post-reform Chinese Cities: A Case Study of Shanghai." *Transactions of the Institute of British Geographers* 33(3): 404–419.

Lin, George C. S. 2007. "Reproducing Spaces of Chinese Urbanisation: New City-based and Land-centred Urban Transformation." *Urban Studies* 44: 1827–1855.

Liu, Ye, Zhigang Li, and Werner Breitung. 2012. "The Social Networks of New-generation Migrants in China's Urbanized Villages: A Case Study of Guangzhou." *Habitat International* 36(1): 192–200.

Liu, Yuting, Shenjing He, Fulong Wu, and Chris Webster. 2010. "Urban Villages under China's Rapid Urbanization: Unregulated Assets and Transitional Neighbourhoods." *Habitat International* 34(2): 135–144.

Ma, Laurence J. C. 2005. "Urban Administrative Restructuring: Changing Scale Relations and Local Economic Development in China." *Political Geography* 24: 477–497.

O'Brien, Kevin J., and Rachel E. Stern. 2008. *Studying Contention in Contemporary China.* Cambridge, MA: Harvard University Press.

Oi, Jean C. 1992. "Fiscal Reform and the Economic Foundations of Local State Corporatism in China." *World Politics* 45: 99–126.

Oi, Jean C. 1995. "The Role of the Local State in China's Transitional Economy." *The China Quarterly* 144: 1132–1149.

Pei, Minxin. 2006. *China's Trapped Transition: The Limits of Developmental Autocracy*, Cambridge: Harvard University Press.

Perry, Elizabeth J. 2008. "Chinese Conceptions of Rights: From Mencius to Mao and Now." *Perspectives on Politics* 6(1): 37–50.

Pow, Choon-Piew 2009. *Gated Communities in China: Class, Privilege and the Moral Politics of the Good Life*. London: Routledge.

Qian, Junxi. 2014. "Deciphering the Prevalence of Neighborhood Enclosure amidst Post-1949 Chinese Cities." *Journal of Planning Literature* 29(1): 3–19.

Read, Benjamin L. 2003. "Democratizing the Neighbourhood? New Private Housing and Home-Owner Self-Organization in Urban China." *The China Journal* (49): 31–59.

Read, Benjamin L. 2008. "Assessing Variation in Civil Society Organizations: China's Homeowner Associations in Comparative Perspective." *Comparative Political Studies* 41(9): 1240–1265.

"Ren Zhiqiang tong pi xiao chanquanfang zhuanzheng: fanzui hefa" Ren Zhiqiang Broadsides Legalizing Small Property-right Houisng, Which Amounts to Legitimating Crimes," sina.com, November 22. Accessed February 1, 2015. http://finance.sina.com.cn/china /20131122/155817408175.shtml.

Shen, Jianfa. 1999. "Modelling Regional Migration in China: Estimation and Decomposition." *Environment and Planning A* 31(7): 1223–1238.

State Council, People's Republic of China. 2015. "Guoxinban jiu 2014 nian guomin jingji yunhang qingkuang juxing fabuhui" (State Council Press Office Issues News Release of Report of the National Economy in 2014), http://www.china.com.cn/zhibo/2015-01/20/content_34593748. htm?show=t, Accessed January 20, 2015.

Tao, Li, Francis K. W. Wong, and Eddie C. M. Hui. 2014. "Residential Satisfaction of Migrant Workers in China: A Case Study of Shenzhen." *Habitat International* 42: 193–202.

Tomba, Luigi. 2005. "Residential Space and Collective Interest Formation in Beijing's Housing Disputes." *The China Quarterly* 184: 934–951.

Tomba, Luigi. 2009. "Of Quality, Harmony, and Community: Civilization and the Middle Class in Urban China." *Positions: East Asia Cultures Critique* 17(3): 592–616.

Tsui, Kai Yuen, 2011. "China's Infrastructure Investment Boom and Local Debt Crisis." *Eurasian Geography and Economics* 52: 686–711

Wang, Fang, and Kam Wing Chan, 2014. "Beijing 'daduhuiqu' de jieding yu tantao" (Discussion on the delimitation of Beijing Metropolitan Area). *Zhongguo kexue (Population Science of China)*. 3: 43–52.

Wang Ya Ping, Humin Du, and Si-ming Li. 2014. "Migrants and the Dynamics of Informal Housing in China. In: *Housing Inequalities in Chinese Cities*, edited by Youqing Huang and Si-ming Li, 87–102. London: Routledge.

Wang, Wenfei Winnie, and Cindy C. Fan. 2006. "Success or Failure: Selectivity and Reasons of Return Migration in Sichuan and Anhui, China." *Environment & Planning A* 38(5): 939–958.

Wang, Wenfei Winnie, and Cindy C. Fan. 2012. "Migrant Workers' Integration in Urban China: Experiences in Employment, Social Adaptation, and Self-Identity." *Eurasian Geography and Economics* 53(6): 731–749.

Wong Keung, Fu, D, Chang Ying Li, and He Xue Song. 2007. "Rural Migrant Workers in Urban China: Living a Marginalised Life." *International Journal of Social Welfare* 16(1): 32–40.

Wu, Fulong. 2002. "China's Changing Urban Governance in the Transition towards a more Market-Oriented Economy." *Urban Studies* 39(7): 1071–1093.

Wu, Fulong. 2004. "Transplanting Cityscapes: The Use of Imagined Globalization in Housing Commodification in Beijing." *Area* 36(3): 227–234.

Wu, Fulong. 2012. "Neighborhood Attachment, Social Participation, and Willingness to Stay in China's Low-Income Communities." *Urban Affairs Review* 48(4): 547–570.

Wu, Weiping. 2004. "Source of Migrant Housing Disadvantages in Urban China." *Environment & Planning A* 26: 1285–1534.

Wu, Weiping, and Wang Guixin. 2014. "Together but Unequal: Citizenship Rights for Migrants and Locals in Urban China." *Urban Affairs Review* 50(6): 781–805.

Wu, Weiping. 2008. "Migrant Settlement and Spatial Distribution in Metropolitan Shanghai." *The Professional Geographer* 60(1): 101–120.

Xu, Jiang, Anthony C. O. Yeh, and Fulong Wu. 2009. "Land Commodification: New Land Development and Politics in China since the Late 1990s." *International Journal of Urban and Regional Research* 33: 890–913.

Xu, Miao, and Zhen Yang. 2009. "Design History of China's Gated Cities and Neighbourhoods: Prototype and Evolution." *Urban Design International* 14(2): 99–117.

Yip, Ngai-Ming, and Yihong Jiang. 2011. "Homeowners United: The Attempt to Create Lateral Networks of Homeowners' Associations in Urban China." *Journal of Contemporary China* 20(72): 735–750.

Yue, Zhongshan, Shuzhuo Li, Marcus W. Feldman, and Haifeng Du. 2010. "Floating Choices: A Generational Perspective on Intentions of Rural – Urban Migrants in China." *Environment & Planning A* 42(3): 545–562.

Yue, Zhongshan, Shuzhuo Li, Xiaoyi Jin, and Marcus W. Feldman. 2013. "The Role of Social Networks in the Integration of Chinese Rural-Urban Migrants: A Migrant-Resident Tie Perspective." *Urban Studies* 50(9): 1704–1723.

Zhu, Jieming. 1999. "Local Growth Coalition: The Context and Implications of China's Gradualist Land Reform." *International Journal of Urban and Regional Research* 23 (3): 534–48.

Instability of migrant labor supply in China: evidence from source areas for 1987–2008

Yan Yuan[a], Zhao Rong[a], Rudai Yang[b] and Liu Yang[c]

[a]Research Institute of Economics and Management (RIEM), Southwestern University of Finance and Economics (SWUFE), Chengdu, Sichuan, China; [b]School of Economics, Peking University, Beijing, China; [c]Nationstar Mortgage Inc., Dallas, TX, USA

Specialists on agricultural economics and economic development of China examine the dynamics of rural-to-urban migration in China from the perspective of migrant labor supply for the period of 1987–2008. Restricted by the *hukou* (registered permanent residence) system, rural households in China usually do not relocate from countryside to cities; only migrant workers travel back and forth. This circular pattern of migrant workers negatively influences their migration stability, thereby resulting in the instability of China's migrant labor supply. By using panel data from rural household surveys in China, we examine migration persistence at both the household level and the individual level. We find that the extent of migration persistence was relatively low in China, suggesting that migrant labor supply was unstable. In addition, we find that labor counts, education level, and network effects improved households' migration persistence. We also find that migration persistence was associated with individual characteristics, such as age, gender, and health conditions.

Introduction

At the beginning of its economic reform, there is no doubt that China was a labor surplus economy. Its huge rural population has provided a seemingly endless supply of cheap labor for the world economy since then. However, in early 2010, it was reported that factories in some coastal cities have been struggling to find workers to meet their export orders. It is puzzling, because only a year prior in 2009 a large number of migrant workers were laid off in coastal cities due to the negative impact of the global financial crisis, as documented by Cai and Chan (2009). In the broad picture of China's migration history, the phenomenon of labor shortages was nothing new: labor shortages occurred in coastal areas as early as in 2003 (Chen and Hamori 2009). These occasionally occurring labor shortages seem controversial – even now, there is still a labor surplus in rural areas (Kwan 2009; Minami and Ma 2009), as there was in 2003.

It is hard to reconcile the coexistence of labor shortages and labor surplus in a static aspect when migration is regarded as permanent.[1] Chan (2010) insightfully argues that one can explain this paradox when taking into account China's socioeconomic contexts. Restricted by the *hukou* (registered permanent residence) system, rural households in China usually do not relocate from countryside to cities; only migrant workers travel back and forth. The labor shortages in 2010 become intuitive when one takes the circular nature of migration into account: migrant workers who experienced massive layoffs in 2009 rationally lowered their expectations on the chance of finding a job. Once they returned home, they were less likely to migrate again based on their adjusted expectations; consequently, this resulted in labor shortages in the coming year (2010) when foreign demand suddenly increased. Overall, the existence of labor shortages highlights the importance of examining the circular nature of China's rural-to-urban migration.

Moreover, the temporariness of rural-to-urban migration is somewhat "permanent," as pointed out by Chan (2001). Due to the *hukou* system, it is extremely hard for ordinary rural households to convert their rural *hukou* to urban *hukou* (2001). On the one hand, a lack of job opportunities in source areas persistently pushes rural laborers to migrate to urban areas. On the other hand, even after working for many years, these migrant laborers can hardly settle down in urban areas because of the *hukou* restriction. Moreover, they are discriminated against in accessing social services and welfare benefits such as children's access to local public schools, social insurance, and pension plans. (Chan 2012). Consequently, without urban *hukou*, these migrant laborers have become a large and highly mobile workforce for China's economy (Chan 2010).

The circular migration also implies that, once back home, migrant workers may change their mind and choose to stay at home rather than migrate again. This migration instability induced by the *hukou* system, from the aspect of destination areas, well reflects the instability of China's migrant labor supply. Given the large number of migration workers in China, the associated costs to the economy, which include the cost of occasional labor shortages, should be substantial. Consider a factory that relies solely on migrant workers. If the likelihood that a migrant worker does not come back the next year is very high, say 50 percent, then the factory will have to bear the burden of at least 50 percent turnover at an annual basis.

Despite its importance, the circular nature of China's migration is so far understudied. The lack of attention is partially due to the convention of the mainstream migration, which is primarily in permanent migration literature,[2] and also partially because of the difficulty in obtaining appropriate data. Since internal migration in most countries is characterized as permanent, studies on determinants of internal migration are reasonably based on cross-sectional data.[3] There has been an array of papers on the determinants of migration in China (e.g. Chen and Hamori 2009; Knight, Deng, and Li 2011; Liang and Ma 2004; Zhao 1999) based on cross-sectional data. Though it makes sense when permanent migration is examined, cross-sectional evidence can hardly make convincing references about the circular nature of China's migration. As argued by Hsiao (1986), it is difficult and unconvincing to make inferences about dynamics of change from cross-sectional evidence.

We are able to overcome the data difficulty by using micro-level panel data. Our data come from the annual household surveys conducted by the Survey Department of the Research Center on the Rural Economy (RCRE), which cover the period from 1986 to 2008. The survey is a longitudinal survey through tracking the same households over years. Particularly, the panel provides the dynamics of a household's migration status, which fit well with our research purpose of investigating China's migration stability. Our data reveal that the proportion of migrant households increased dramatically, from 12 percent in 1987 to 66 percent in 2008. The examination period thus provides us with enough migration variations over time.

A household can be either migrant or non-migrant in a given year. Accordingly, we classify two types of migrant households: inexperienced migration (referring to those household-years that are migrant in the current year but not in the previous year) and experienced migration (referring to those household-years that are migrant in both years). Our focus is on the propensity of experienced migration; that is, the likelihood of migration among experienced household-years (household-years that are migrant in the previous year). We use the propensity to measure the extent of migration stability. Meanwhile, our data reveal that the proportion of experienced migration among the total migration dominated in the later years, indicating experienced migration has become the major force of migrant labor supply. This further rationalizes our focus on the propensity of experienced migration.

We first estimate the likelihood of migration among experienced households for the sub-period 1987–1999. By examining rural households in Sichuan Province, Zhao (1999) finds that migration barriers tended to stimulate better-educated rural laborers to migrate. Our study gains another deeper insight: education, as well as labor surplus and network effects, not only encouraged inexperienced households to take the first step and become migrant, but also improved their migration stability. We also find that the propensities of experienced migration, though increasing, varied over time. These propensities also varied across regions, and were strongly influenced by local migration regulations.

The availability of individual information since 2003 enables us to estimate the propensity of experienced migration both at the household level and at the individual level for the sub-period 2004–2008. It turns out that the results of both regressions are similar. Now that individual characteristics are available, we further examine how individual characteristics were related to the extent of migration stability. We find that younger laborers, male laborers, or laborers with better health conditions were more likely to keep on migrating.

The rest of the paper is constructed as follows. Section two discuss the research question and analytical framework. Section three describes the data and variables. Section four reports estimation results. Section five concludes.

Research question and model specification

Research question

Since economic reform in 1978, China has experienced rapid structural changes and remarkable transformations, marked by reduced state control over labor mobility. It has been well acknowledged that there has been a steady flow of labor from agriculture to industry and from rural areas to urban areas (Cai, Park, and Zhao 2008; Chang, Dong, and MacPhail 2011). The stock of migrant labor has increased from about 30 million in the late 1980s to between 150 and 180 million in recent years (Fan 2009; Liang and Ma 2004). Our sample confirms the expansion at the household level. Figure 1 presents the proportion of migrant households over years from 1987 to 2008. As shown, the migration proportion was only 12 percent in 1987, and it reached 66 percent in 2008. The growth rate between 1999 and 2004 was the fastest, perhaps due to China's entry to the WTO. Unfortunately, data limitation prevents us from going further in this direction. Figure 1 also shows that the growth accelerated after 1995.[4]

One important factor that drove the migration wave was, of course, lack of job opportunities in source areas, which was reflected as rural labor surplus. Even by now,

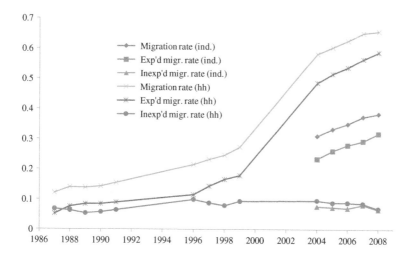

Figure 1. Proportions of inexperienced and experienced migration, 1987–2008.

studies still find that there exists labor surplus in rural areas (Golley and Meng 2012; Kwan 2009; Minami and Ma 2009). Though there is some evidence of rising nominal unskilled wages in cities, Golley and Meng (2012), by using the RUMiC Rural House-hold Survey data, find no evidence that the wage increase was caused by unskilled labor shortages. They also demonstrate that by 2008 there were still abundant rural laborers with earnings far below their migrant counterparts, suggesting that they were likely underemployed.

Despite the surge in migration, there were few institutional arrangements to help migrant workers settle down in destination areas. Specifically, an institution, the *hukou* system,[5] which was used to sanction rural-to-urban migration in the planned-economy era, is still fully in force today. This institution makes it extremely hard for ordinary rural households to obtain permanent residency in cities. Chan and Buckingham (2008) document the policy change of permanent rural-to-urban migration (*nongzhuanfei*, or converting rural *hukou* to urban *hukou*). Empirical evidence shows that it is rare that migrant laborers obtain urban *hukou* in the destination city and finally settle down as permanent migrants (Hu, Xu, and Chen 2011). As a result, it makes internal migration in China unique to most other developing countries: instead of relocating permanently from countryside to cities, rural households usually choose to stay in the countryside; only their migrant workers travel back and forth.

Moreover, the above circular pattern tends to be prevalent and long term because of the institutional restrictions. Migrant workers may return home for various reasons, such as family issues or unsuccessful work experiences. However, lack of job opportunities in rural areas would later push them to migrate again. Thus, with the expansion of the migrant labor pool, the proportion of migrant laborers with migration experiences would increase. While this component change in the migration pool has been noted in the liter-ature, it is not well examined, perhaps due to data limitations.

The panel data enable us to thoroughly investigate the composition of migrant households and its evolution over time. We categorize two types of migrant households: inexperienced migrant households (migrant household-years with less than or equal to 180 migrant labor-days in the previous year) and experienced migrant households (migrant household-years with more than 180 migrant labor-days in the previous year). Figure 1 also presents the proportions of inexperienced and experienced migrant households. Consistent with our expectation, the experienced migration rate increased dramatically from 5.3 percent in 1987 to 59 percent in 2008. Its growth was mild in the early years, and accelerated in late 1990s. In contrast, there was no clear uptrend of the inexperienced migration rate. It suggests that the growth of the experienced migration rate was the major contributor to the growth of the migration rate.

As to the component change, the proportion of experienced migrant households among all migrant households increased from 43 percent in 1987 to 90 percent in 2008; accordingly, the proportion of inexperienced migrant households decreased from 57 percent to only 10 percent. Overall, the diffusion of migration in China was characterized by a significant change in the composition of migrant labor supply: in the early years, migrant labor supply relied heavily on inexperienced migration; in the later years, the importance of inexperienced migration became minor, and experienced migration became the major source. Given the dominant proportion of experienced migration among the total migration in the later years, experienced migration should have become the major force that drives the stability of migrant labor supply. This further rationalizes our focus on examining migration propensities among experienced households.

To have some idea about how the migration stability evolved over time, we present the migration propensity among experienced households in Figure 2. As shown, the propensity of experienced migration had an uptrend over time. It was around 60 percent in the late 1980s, began to increase in the late 1990s, then was stabilized at the level of near 90 percent since 2004. We also present the migration propensity among inexperienced households in Figure 2. As shown, the propensity of inexperienced migration remained at a low level of around 10–20 percent.

The baseline model

Theoretically, disparities in regional economic opportunities and geographic amenities are regarded as the most important determinants of migration (Borjas 1999; Massey

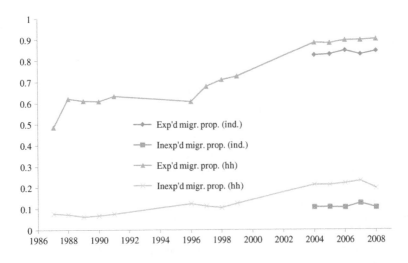

Figure 2.   Propensities of inexperienced and experienced migration, 1987–2008.

et al. 1993; Todaro 1969, 1976). We follow a generalized Todaro paradigm (Todaro 1976). We postulate that households make rational choices: they choose to send out migrant laborers when the expected earnings or welfare are higher than having them stay at home. We assume that a rural household allocates its total labor supply between local production activities and migration activities to maximize its total income. A household makes its migration choice at the beginning of the year based on information acquired from both the previous and current year. The reduced-form equation of the net gain from migration is described as[6]

$$dY_{i,t} = Y_{i,t}^1 - Y_{i,t}^0 = X_{i,t}\beta + U_{i,t}^1 - U_{i,t}^0 \tag{1}$$

$$dY_{i,t} > 0 \quad \text{iff} \quad D_{i,t} = 1$$

$$dY_{i,t} \leq 0 \quad \text{iff} \quad D_{i,t} = 0.$$

$Y_{i,t}^1$ denotes household i's total income (net of migration costs) in year t in the presence of migrant labor, and $Y_{i,t}^0$ denotes household total income in the absence of migrant labor. $dY_{i,t}$ is a latent variable denoting the net gain from migration. $D_{i,t}$ is an observable discrete choice variable indicating whether household i contains migrant labor in year t; it equals one if so, zero otherwise. $D_{i,t} = 1$ implies that it is profitable for household i to become a migrant household in year t while $D_{i,t} = 0$ implies that it is profitable to employ all labor at home. $X_{i,t}$ is a vector of household and village characteristics that affect the net gain $dY_{i,t}$.

Regarding a household's information about destination areas, its migration status in the previous year plays an important role. For inexperienced households, lack of relevant information should be a major barrier to their first-time migration. In contrast, for experienced households, the information problem should be significantly reduced since they already have access to the job market and have been more integrated in the destination (Farre and Fasani 2011). Therefore, households' decision on whether to send out migrant labor should be made differently when their previous migration status is different. Additionally, as revealed by our data, the probability for an experienced household to be migrant was 83 percent whereas this probability was only 9 percent for an inexperienced household. This substantial difference in migration propensities strongly suggests that pooling these two types of households when examining the determinants of migration would lead to biased estimates. Overall, both the theoretical analysis and the summary statistics suggest that we should examine experienced migration and inexperienced migration separately. Additionally, we control for a household's other characteristics in both the current and previous years which influence its migration decision.

As we have discussed, our focus is on the propensity of experienced migration, which should well reflect the stability of migrant labor supply. Given that we are trying to relate the stability to the phenomenon of labor shortages, one may be concerned that the instability of migrant labor supply is driven by the instability of labor demand from destination areas, and such instability should not cause severe labor shortage problems. This concern, though reasonable, relies on the assumption that migrant labor supply responds to demand shocks in a timely manner. Though firms can easily lay off their migrant workers when demand falls, to attract them back is not as easy when demand later increases, given that these workers may have already returned to source areas, which are far away from the destination area.

Moreover, even with perfect information about job opportunities, returned migrants may not react to a demand increase in a timely manner. Upon labor shortages, firms in destination areas should have strong incentives to spread the information, especially to those migrant workers that they have laid off. The information may quickly reach those migrant workers who have returned home, thanks to the popularity of cell phones in China. However, there are several factors that prevent these returned migrants from responding promptly. First, some of them may have already found a new job, and the labor contract may not allow them to quit immediately even if the potential job is more attractive. Additionally, the cost of migration, family issues, and uncertainties involved can further prevent returned migrants from a prompt response. Meanwhile, it is difficult to make up for labor shortages in the destination area by other sources, e.g. inexperienced households in faraway rural areas are subject to the information problem while labor surplus in nearby rural areas may have been already exhausted. Consequently, even if the instability of migrant labor supply is solely driven by demand shocks from destination areas and the information on job opportunities is perfect to returned migrants, it can still result in occasional labor shortages, such as the one in 2010.

To quantitatively analyze the propensity of experienced migration, we use the following linear probability specification with the sample restrict to experienced household-years:

$$\text{Prob}(D_{i,t} = 1) = X_{i,t}\beta_1 + \text{Year\_dummy} + \text{Prov\_dummy} + \varepsilon_{i,t}, \qquad (2)$$

where $D_{i,t}$ is the household migration dummy, indicating whether household i has migrant labor-days greater than 180 in year t; it equals one if so, and zero otherwise. Our major interest is in the time variation of the stability of migrant labor supply, so the variables of interest are year dummies. Year dummies capture the effects of time-variant macro characteristics that influence households' time-variant average propensity to migrate. To investigate the time trend and volatility of migration propensities, it is better to estimate without control for household fixed effects. If household dummies are included, the cross-sectional variations that we are interested in would be removed. To control for time-invariant effects at the province level, we include province dummies.

The survey data at the individual level also enable us to investigate migration persistence by examining migration propensities among experienced laborers. We estimate the following specification with the sample restricted to experienced laborers.

$$\text{Prob}(D_{j,t} = 1) = Z_{j,t}\beta_2 + \text{Year\_dummy} + \text{Prov\_dummy} + \varepsilon_{j,t}, \qquad (3)$$

where $D_{j,t}$ is the individual migration dummy, indicating whether laborer j has migrant labor-days greater than 180 in year t; it equals one if so, and zero otherwise. $Z_{j,t}$ is a vector of individual, household, and village characteristics.

Data and variables

Our data come from annual household surveys conducted by the Survey Department of the Research Center on the Rural Economy (RCRE). The RCRE survey was originally designed as a longitudinal survey through tracking the same households over years. Our data cover the period of 1986–2008, with interruptions of 1992, 1994 and 2000–2002. Due to lack of information on migration in the 1993 survey, we only use panel data of 17 years in three periods: 1986–1991, 1995–1999, and 2003–2008. In practice, we pool the first two periods and treat them separately from the last period. In the 86–99 data, we have information for the following six provinces: Zhejiang, Guangdong, Hunan, Sichuan, Jilin, and Gansu. The 03–08 data are retrieved from the survey separately, with five additional provinces included. They are Shanxi, Jiangxi, Fujian, Henan, and Hubei. The actual number of villages surveyed each year is determined by the size of the province. Before 2000, on average 64 villages are surveyed, covering about 3500 households; since 2003, the number increases to 85 villages, covering 5700 households.

Regarding the representativeness of the sample, we restrict our discussion to the 03–08 survey data, which are more up to date and better represent the current situation. First of all, instead of only having six provinces as in the 86–99 data, the 03–08 data comprise five additional provinces, leading to 11 provinces included. These provinces well represent different regions and economic conditions in China. Because of this data expansion, the representativeness should have been naturally improved. Second, The RCRE survey itself to some extent has represented the situation of the nation. As argued by Benjamin, Brandt, and Giles (2005),

> in each province, counties in the upper, middle and lower income terciles were selected, from which a representative village was then chosen. Subject to the limits of this stratification, the RCRE sample should reasonably capture both inter- and intra-provincial income variation.

We further compare the 03–08 data of 11 provinces with the nationwide RCRE survey data with respect to the means of some major indicators, and our sample of 11 provinces is not significantly different from the full sample of the survey. We also examine the representativeness of our data by comparing our 2005 individual sample with the 2005 Census (1 percent sample) data. We find that the age distribution and the education distribution are close between these two samples. Overall, the comparisons suggest that our 03–08 data, to some extent, represent the situation of the whole nation. Given the differences revealed from the comparisons, we admit that one should be cautious when applying the conclusion in this paper and expand it to the whole nation. The complete comparisons are presented in Appendix 2.

Making this period more interesting, it includes the period when circular migration was for the first time legally accepted since the establishment of the *hukou* system. In 1984, the government started to allow peasants to stay in towns contingent on that they could be self-sufficient, i.e. running a business or working in Township and Village Enterprises (*xiangzhen qiye*). In 1985, the government began to sanction the Temporary Residence Permit (*zanzhuzheng*) to rural-to-urban migrants. These two policies acknowledged for the first time the legal existence of temporary residency in cities and promoted rural-to-urban migration afterwards. Thanks to reduced state control over labor mobility, it has been witnessed a steady flow of labor from rural to urban areas. According to the National Bureau of Statistics (NBS), there were 168 million rural-to-urban migrants in 2014 (National Bureau of Statistics 2014).

The survey itself does not provide a consistent definition of migrant households over the survey period. From 1986 to 1991, a migrant household can be defined as one that has at least one migrant worker. However, the information about the number of migrant workers is no longer reported after 1991. Meanwhile, migrant labor-days are persistently reported in all years. For years when the number of migrant workers is reported, the average migrant labor-days per migrant worker are close to six months. In addition, the 180-day cutoff is the criterion used by the census to define residents, and it has the advantage of assigning a person to one and only one place in any 12-month period. We therefore use migrant labor-days to define a migrant household and choose 180 as the threshold. The threshold of about six months is also consistent with the definition of "temporary population" (*zanzhu renkou*) by the NBS (Chan 2013).[7] Specifically, we define a "household-year" as migrant when its total migrant labor supply is greater than 180 labor-days in the year. A household-year is defined as non-migrant when its total migrant supply is equal to or below 180 labor-days. Accordingly, we generate a migration dummy to indicate whether a household-year is migrant; it equals one if so, and zero otherwise.

This is how we define whether a household is migrant or not for the years before 2000. Since 2003 the survey began to collect individual information instead of only asking questions at the household level. Regarding the migration information, only migrant labor-days aggregated at the household level are reported in the previous years; since 2003 these migrant labor-days are reported at the individual level. Accordingly, we generate the migration dummy at the household level based on the aggregated migrant labor-days of all family members in the household.

The limited information of individual migration only allows us to use the household as the major decision-making unit. Nevertheless, this choice is acceptable for two reasons. First, rural economic structures in China exhibit strong family ties (Wang and Zuo 1999). Second, evidence shows that in rural China most economic decisions, including the migration decision, are undertaken to maximize the welfare of the entire family (Zhao 1999). Additionally, taking advantage of the individual information available since 2003, we also investigate migration dynamics at the individual level by using the 03–08 data.

To clean the data, we delete the following types of observations from the original data-set (72,420 observations). First, all variables are lagged by one year except for the migration dummy, average schooling years, and labor counts. Observations in 1986, 1995, and 2003 are dropped because of the lag specification. Second, we delete observations with labor counts equal to zero. Third, we delete observations that have at least one key variable with invalid values. Key variables refer to the following variables: the one-year lagged migration dummy, agricultural land size, taxes, village fees, the current migration dummy, average schooling years, and labor counts. Last, we delete observations that have at least one key variable either less than the 0.5 percentile or greater than the 99.5 percentile.[8] From this procedure, we end up with a total sample of 54,243 observations. Appendix 1 presents how the sample size is reduced from 72,420 to 54,482 observations.

Table 1 reports the summary statistics of household characteristics. The monetary terms are in real value (1987 = 100), calculated based on the Rural Consumer Price Index. All variables are one-year lagged except that the migration dummy, average schooling, and labor counts are current. Overall, 38 percent of household-years were migrant during the examination period.

One important measure of human capital is the education level. In the data-set, education status takes four different values. Level 1 corresponds to illiterate or semi-illiterate, level 2 to primary school, level 3 to junior high school, and level 4 to senior high school and above. Using the terminology of China's education system, illiterate or semi-literate corresponds to 0 years in school, primary school to six years in school, and junior high school to nine years in school. Because the data do not release education levels above senior high school, all education levels of senior high school and above are coded as 12 years in school. To capture the overall human capital of a household,

Table 1. Summary characteristics of rural households and labor.

Examination level:	Household						Labor	
	Full sample		87–99 (6 provinces)		04–08 (11 provinces)		04–08 (11 provinces)	
Sample:	Mean	SD	Mean	SD	Mean	SD	Mean	SD
Migration dummy (yes = 1)	0.38	0.49	0.18	0.38	0.63	0.48	0.34	0.48
Lagged migration dummy (yes = 1)	0.36	0.48	0.17	0.37	0.61	0.49	0.32	0.47
Schooling (years)	6.92	2.30	6.47	2.48	7.50	1.90	7.51	2.82
Labor counts (persons)	2.83	1.15	2.68	1.13	3.01	1.16	3.45	1.17
Lagged landholding (mu)	8.12	8.53	9.70	8.99	6.13	7.44	6.31	7.47
Lagged taxes (yuan)	167.35	720.71	96.32	208.56	256.81	1050.92	286.18	1204.43
Lagged village fees (yuan)	134.14	626.33	74.65	108.22	209.07	928.17	216.67	951.73
Lagged village migration rates	0.34	0.27	0.16	0.16	0.56	0.20	0.32	0.13
Age (years)							37.44	12.36
16–30							0.35	0.48
31–45							0.34	0.47
46–60							0.32	0.47
Being male (yes = 1)							0.55	0.50
Having a child (yes = 1)							0.22	0.41
In poor health condition (yes = 1)							0.08	0.27
Observations	54,243		30,234		24,009		64,293	

average schooling is defined as the average of school years of all laborers. The average schooling was 6.9 years, slightly above the level of primary school.

On average, a rural household had 2.8 laborers, cultivated a plot of 8.1 mu (1 mu = .07 hectare or .16 acre). Taxes paid were 167 yuan and village fees were 134 yuan annually per household. The real burden of taxes and fees imposed upon a rural household could be heavier. Hidden taxes may take different forms, such as compulsory grain sales to the government at lower-than-market prices, free labor provision for public construction projects, etc. Nevertheless, these two measures provide a lower bound for taxes paid by an average rural household.

In practice, we regress among the 87–99 subsample and the 04–08 subsample separately. We thus present their summary statistics in Table 1, respectively. There are 30,234 observations representing 5208 households for the 87–99 subsample, and 24,009 observations representing 5570 households for the 04–08 subsample.

We then divide the sample into two groups, inexperienced households (household-years with the lagged migration dummy equal to zero) and experienced households (those with it equal to one). For our research purpose, Table 2 presents the summary characteristics of experienced households by their current migration status. As shown, given that a household had migrated in the previous year, its probability of being migrant in the current year was 83 percent (16,264/(3339 + 16,264)). Alternatively speaking, there were on average 17 percent experienced households that stopped migrating in the current year, suggesting a high degree of instability of migrant labor supply. We also report the 87–99 subsample and the 04–08 subsample, respectively. In either period, compared to those that stopped migrating, households that kept on migrating had higher average schooling, more laborers, less agricultural land, and were faced with higher village migration rates.

Changes of migration propensities over time

To understand how experienced migration propensities varied across provinces, Table 3 presents the average propensities of experienced migration in three sub-periods (1987–1991, 1996–1999, and 2004–2008) for six provinces. In general, compared to the period of 1987–1991, experienced migration propensities were higher in 1996–1999, and were the highest in 2004–2008, which is consistent with the pattern in Figure 2.

In 1987–1991, the propensity of experienced migration in Guangdong was 77 percent, the highest among the six provinces. It is intuitive, given that Guangdong was the richest among the six provinces and most of its migrant labor stayed within the province. Similarly, Zhejiang, the second richest, experienced the second highest propensity at 58 percent in 1987–1991. The propensities were less than 50 percent for the rest four provinces, which were relatively poor.

In contrast, in terms of the growth of experienced migration propensities, the data reveal a different picture. The highest growth happened in Hunan, an inland province contiguous to Guangdong: its average propensity in 1987–1991 was 42 percent and reached 69 percent in 1996–1999. Sichuan, another inland province, has the second highest growth from 46 to 68 percent. It is somewhat surprising not to witness similar growth in Jilin and Gansu. The growth in either Zhejiang or Guangdong was mild, indicating that the migration diffusion in either province was approaching some saturation level. Overall, the enhanced stability of migrant labor supply should be attributed to the growth of experienced migration propensities in Hunan and Sichuan, which sent out a

Table 2. Characteristics of experienced households and experienced labor.

Examination level:	Experienced households						Experienced labor	
	Full sample		87–99 (6 provinces)		04–08 (11 provinces)		04–08 (11 provinces)	
Sample:	Non-migrant	Migrant	Non-migrant	Migrant	Non-migrant	Migrant	Non-migrant	Migrant
Migration dummy	0.00	1.00	0.00	1.00	1.00	1.00	0.00	1.00
	(0.00)	(0.00)	(0.00)	(0.00)	(0.00)	(0.00)	(0.00)	(0.00)
Lagged migration dummy	1.00	1.00	1.00	1.00	0.00	1.00	1.00	1.00
	(0.00)	(0.00)	(0.00)	(0.00)	(0.00)	(0.00)	(0.00)	(0.00)
Schooling (years)	6.89	7.50	6.46	6.87	7.38	7.66	7.75	8.18
	(2.25)	(1.85)	(2.37)	(2.05)	(1.98)	(1.77)	(2.50)	(2.41)
Labor counts	2.88	3.30	2.94	3.39	2.82	3.27	3.49	3.48
	(1.14)	(1.18)	(1.16)	(1.31)	(1.10)	(1.14)	(1.17)	(1.20)
Lagged landholding (mu)	7.33	5.93	8.42	7.73	6.07	5.48	5.33	5.06
	(7.68)	(6.33)	(8.20)	(6.99)	(6.81)	(6.07)	(6.04)	(5.93)
Lagged taxes (yuan)	162.44	244.11	90.43	92.93	246.35	281.78	275.53	327.67
	(716.24)	(1061.23)	(218.14)	(163.05)	(1021.12)	(1180.30)	(1247.23)	(1339.86)
Lagged village fees (yuan)	131.71	196.97	63.40	67.11	211.31	229.32	214.13	276.82
	(731.51)	(834.24)	(96.08)	(105.74)	(1066.10)	(928.08)	(1000.37)	(1060.64)
Lagged village migrant rates	0.38	0.58	0.23	0.34	0.56	0.63	0.34	0.38
	(0.24)	(0.21)	(0.16)	(0.22)	(0.18)	(0.16)	(0.12)	(0.12)
Age (years)							34.64	32.58
							(11.44)	(10.25)
16–30							0.44	0.50
							(0.50)	
31–45							(0.50)	0.35	0.36
							(0.48)	
46–60							(0.48)	0.21	0.14
							(0.41)	(0.34)
Being male (yes = 1)							0.61	0.65
							(0.49)	(0.48)
Having a child (yes = 1)							0.25	0.24
							(0.43)	(0.43)
In poor health condition (yes = 1)							0.06	0.04
							(0.24)	(0.20)
Observations	3339	16,264	1797	3244	1542	13,020	3445	17,237

Note: Standard deviations are in parentheses.

Table 3. Migration propensities over time periods and cross provinces.

| | Inexperienced migration propensity | Experienced migration propensity | | | | | | |
	Full sample	Full sample	Zhejiang	Guangdong	Hunan	Sichuan	Jilin	Gansu
Household level:	Migration defined as migrant labor-days greater than 180							
1987–1991	0.07	0.60	0.58	0.77	0.42	0.46	0.46	0.49
1996–1999	0.12	0.68	0.63	0.80	0.69	0.68	0.52	0.36
2004–2008	0.22	0.89	0.89	0.93	0.89	0.90	0.83	0.81
	Migration defined as having one migrant labor							
2004–2008	0.23	0.88	0.89	0.92	0.86	0.89	0.82	0.80
Individual level:	Migration defined as migrant labor-days greater than 180							
2004–2008	0.11	0.83	0.86	0.90	0.80	0.82	0.78	0.76

large number of migrant workers to Guangdong as well as other coastal areas during the study period.

In the third sub-period of 2004–2008, the average propensity reached a new high. Most importantly, its cross-province variation became smaller, as the propensities in all six provinces were over 80 percent. This convergence should be largely attributed to massive migration that happened across provinces, especially from inland provinces to coastal ones. It also indicates that the temporary characteristic of migration still existed (the propensity was not very close to 100 percent), and there were some barriers that systematically prevented the propensity from further increasing. We tend to believe these barriers were largely associated with the *hukou* system.

One may be concerned that using 180 migrant labor days at household level may not reliably reflect the real migration status of the household. We thus use 180 migrant labor-days at the individual level to define whether an individual is migrant, and then re-define a migrant household as having at least one migrant family member. Data limitation only allows us to do so for the 04–08 sample. We report migration propensities in Table 3 based on the alternative definition. As shown, these migration propensities are similar to those using the original definition. Therefore, the estimation bias, if any, should be minor based on the original definition.

Overall, we tend to believe that the instability of migrant labor has been improved over time, but remained substantial, and the major blame was most likely laid on the unchanged *hukou* system.

Empirical results

In the following two subsections, we first examine the migration dynamics at the household level for the periods 1987–1999 and 2004–2008, respectively. Then, we examine the dynamics at the individual level for the period 2004–2008.

Instability of migrant labor supply at the household level

87–99 regressions

We first estimate the probability of migration among inexperienced households for 1987–1999. As shown in column (1) of Table 4, the estimated coefficients on year

Table 4. Changes in migration propensities over time, 1987–1999.

Dependent variable:	Inexperienced migration		Experienced migration				
Sample:	Full sample (1)	Guangdong (2)	Full sample (3)	Zhejiang (4)	Guangdong (5)	Hunan (6)	Sichuan (7)
Year dummy: 1987	-.024***	.057***	-.16***	-.25***	.00041	-.31***	-.25***
	(.009)	(.021)	(.033)	(.075)	(.053)	(.082)	(.095)
Year dummy: 1988	-.028***	.017	-.049*	-.13*	.033	-.13	-.24***
	(.0084)	(.019)	(.028)	(.07)	(.041)	(.087)	(.083)
Year dummy: 1989	-.041***	.028	-.075***	-.098	-.028	-.26***	-.11
	(.0082)	(.019)	(.028)	(.066)	(.041)	(.076)	(.092)
Year dummy: 1990	-.038***	.018	-.079***	-.19***	.043	-.19**	-.23***
	(.0082)	(.019)	(.027)	(.065)	(.038)	(.08)	(.081)
Year dummy: 1991	-.032***	.024	-.05*	-.020	-.0054	-.17**	-.15**
	(.0083)	(.019)	(.026)	(.061)	(.04)	(.077)	(.071)
Year dummy: 1996	.0091	.19***	-.063**	-.064	.11***	-.14**	-.12**
	(.0091)	(.021)	(.025)	(.056)	(.042)	(.061)	(.059)
Year dummy: 1997	.00013	.062***	-.059***	-.14**	-.0063	-.098	.00083
	(.0092)	(.020)	(.023)	(.058)	(.031)	(.062)	(.06)
Year dummy: 1998	-.014	.038*	-.014	-.053	-.027	-.026	.031
	(.0093)	(.021)	(.022)	(.058)	(.031)	(.059)	(.054)
Schooling	.0038***	.0083***	.0094***	.013*	-.0019	.016	.0085
	(.00068)	(.002)	(.0035)	(.0078)	(.005)	(.011)	(.0098)
Labor counts	.04***	.038***	.067***	.08***	.039***	.096***	.11***
	(.0022)	(.0046)	(.0056)	(.014)	(.0074)	(.019)	(.018)
Log (lagged landholding)	-.0058*	.014*	.0029	-.014	.027*	-.042	-.12***
	(.0031)	(.0071)	(.011)	(.021)	(.015)	(.036)	(.045)
Log (lagged taxes)	-.0067***	-.009***	-.014***	-.0061	-.016**	-.026	.0041
	(.0015)	(.0025)	(.0051)	(.0089)	(.0068)	(.019)	(.033)
Log (lagged village fees)	.0021*	.00078	.0067	.00046	.013**	.0027	-.03
	(.0012)	(.0026)	(.0043)	(.0084)	(.0058)	(.013)	(.028)
Lagged village migrant rates	.29***	.32***	.44***	.047	.48***	.73***	.30**
	(.021)	(.041)	(.037)	(.15)	(.045)	(.16)	(.14)
Observations	25,193	4851	5041	1089	1860	755	734
Adjusted R^2	0.053	0.065	0.127	0.049	0.099	0.142	0.101

Note: Standard errors are in parentheses.
*$p < 0.1$; **$p < 0.05$; ***$p < 0.01$.

dummies (the reference year is 1999) are significantly negative in the 87–91 period, indicating that migration propensities among inexperienced households were relatively higher in the 96–99 period. This increase in inexperienced migration propensities should have been driven by the combination of various macrofactors, such as more information about migration available to rural households, more accommodating government policies toward rural migrants, less state control over labor mobility, etc.

As to household-level characteristics, the coefficient on labor counts is positive and significant at the 1 percent level. This is consistent with the consensus that rural labor surplus leads to migration. It also reflects, to some extent, that the migration decision was made at the household level. The coefficient on schooling is also significantly positive. Higher educated individuals may be more adaptive and more capable to learn, thus better prepared to migrate. The coefficient on lagged landholding is negative and significant at the 10 percent level. As to village-level characteristics, the coefficient on lagged taxes is significantly negative, consistent with the argument that tax burdens drive rural households to migrate. The magnitudes of the coefficients indicate that both landholding and taxes played a negligible role. The effect of lagged village fees is somehow positive and significant at the 10 percent level. The effect of lagged village migration rates is significantly positive, suggesting the existence of network effects on the diffusion of migration.[9]

In contrast, the year effects reveal a very different pattern when we estimate the probability of migration among experienced households. As shown in column (3), the coefficients on year dummies (again, the reference year is 1999) are also mostly negative. However, there is no clear pattern that the magnitude of year effects is lower in the early years. As to other characteristics, the coefficients on labor counts, average schooling, lagged taxes, and lagged migration rates share the same sign and significance to those in column (1). Therefore, these factors not only influenced inexperienced households to take the first step and become migrant, but also influenced their migration persistence.

As revealed by the summary statistics, experienced migration propensities varied across regions. One important issue is that the proportions of migrants who traveled out of their home province were different across regions. Specifically, in regions where the economy is more developed such as Zhejiang and Guangdong, migrants are more likely to find jobs within their home provinces, thus their commuting costs are lower. However, for regions where the economy is less developed like Hunan and Sichuan, migrants are more likely to travel out of their home provinces. Indeed, Hunan and Sichuan are among those provinces that send out the largest number of migrant workers. From this aspect, it is also interesting to investigate how experienced migration propensities evolved over years in these labor-sending-out provinces.

We thus repeat the regression of column (3) by provinces in columns (4)–(7). By comparing the coefficients on year dummies over these estimation results, we come up with two major findings as follows. First, there were strong upper trends of experienced migration propensities in the labor-sending-out provinces, Hunan and Sichuan (columns 6 and 7). This is consistent with the conventional observation that the growth of rural-to-urban migration was characterized by the flow-out of migrant workers from inland to coastal areas, or from less-developed areas to developed areas. It is thus not surprising that there were no such rises of experienced migration propensities in developed areas such as Zhejiang and Guangdong (columns 4 and 5). To save space, we do not report the regressions for Jilin and Gansu because the valid sample size is relatively small in these two provinces, with the number only about 300.

Second, the migration policies of destination areas may have influenced migration propensities. Column (2) reruns the regression of column (1) with the sample restricted to Guangdong. It shows that in 1996 the propensity of inexperienced migration in Guangdong was 19 percent higher than the reference year 1999. We trace the migration policy changes in Guangdong, and find that in 1995 the Guangdong Government sanctioned an official document to regulate temporary residents (*liudong renkou guanli banfa*). It required that local governments assign staffs specialized in administering migrant labor, with special efforts being put on birth control (*jihua shengyu*). The enforcement of this policy should have adversely influenced migration propensities in source areas, such as Hunan and Sichuan, in which a large portion of migrant labor flew out to Guangdong.[10] Our estimation results confirm this. As shown in columns (6) and (7), both year effects of 1996 are negative and significant at the 5 percent level, indicating that the propensities of experienced migration in Hunan and Sichuan were adversely influenced in 1996.

As shown, the magnitude of the 1996 year effect is 2 percent higher for Hunan (−0.14) than that for Sichuan (−0.12), indicating that the negative effect of this regulation on Hunan was stronger. It may be because migrant workers from Hunan were more likely to go to Guangdong. Furthermore, the negative effect was somehow time persistent in Hunan: even in the next year 1997, this propensity was still lower than the level of 1999 by about 10 percent, though insignificant.

As a counterfactual test, we note that while the propensity of experienced migration surged in 1996 for Guangdong (column 5), it was not the case for Zhejiang (column 4). Overall, our estimation results indicate that to make up for the reduced labor supply due to the regulation, firms in Guangdong hired more migrant labor (both first time and experienced) from local rural areas. This substitution effect explains why the year effect of 1996 is significantly positive for Guangdong but not for Zhejiang.

In summary, the above analysis provides new insights on the mechanism behind labor shortages by documenting the potential labor supply crisis that happened in Guangdong Province in 1996. Due to the enforcement of Guangdong's migration regulation in 1995, experienced migration propensities in source areas (e.g. Hunan and Sichuan) dropped dramatically in 1996. The reduced labor supply did not turn into labor shortages only because firms in Guangdong managed to make up for this reduction by hiring more migrant workers from local rural areas. This buffer zone due to local labor surplus in Guangdong should have been exhausted since 2003, because it was then that labor shortages appeared in the Zhujiang Delta.

04–08 regressions

We now turn to the 04–08 subsample to explore how volatile the propensity of experienced migration was during this period. We investigate this subsample separately from the 87–99 subsample for three reasons. First, one may be concerned that our examination of migration dynamics at the household level may not well reflect the real situation of migrant labor supply, which should be more accurately measured at the individual level. By running the regressions separately, we are able to address this issue by comparing the 2004–2008 regression at the household level with that at the individual level. Second, the 04–08 data, though from the same source of the 86–99 data, are retrieved separately. There is no guarantee that household identification numbers are consistent with those in the previous years. To minimize potential measurement errors, we treat them separately. Third, the 04–08 subsample is more up to date, thus should better reflect the current situation, which is of special interest to policy makers.

We rerun the regression of experienced migration by using the 04–08 household subsample in column (1) of Table 5. The effects of independent variables are largely similar to those in the regression when the 87–99 subsample is used (column 3 of Table 4). The estimated coefficients on year dummies reveal a narrowed volatility of the propensity: the range of year dummy coefficients is only 2 percent. The result also shows that the experienced migration propensities in 2007 and 2008 were about 2 percent lower than the previous years (the reference year is 2004), perhaps due to the negative shocks from the 2007 financial crisis and the labor law enforcement in 2008. Similar drops are also found in the regression of migration among inexperienced households (unreported).

In summary, household-level regressions show that experienced migration propensities, as the measure of migration persistence, varied over time and across regions. Additionally, the migration persistence was influenced by regional migration regulations. Last, the persistence was stabilized since 2004.

Instability of migrant labor supply at the individual level

The survey began to collect individual information in 2003. The availability of individual information enables us to explore experienced migration in 2004–2008 at the individual level. Now that individual characteristics, such as gender, age, and health condition, are available, we are able to investigate how individual characteristics were related to the migration instability.

Data and summary statistics

The data screening process for the 04–08 individual sample is similar to that for the 04–08 household sample. Unlike the process at the household level, we start with restricting the individual sample to those who are laborers. An individual is defined as a laborer if he/she is not disabled, and his (her) age is between 16 and 60 (16 and 55).[11] For comparison with the household-level regressions, all previously used variables are defined at the household level except that migration status, schooling, and village migration rates are defined at the individual level. A labor-year is defined as migrant if its migrant labor-days exceed 180 days in the year; otherwise, it is non-migrant. As shown in Table 1, we end up with 64,293 observations, representing 18,794 laborers.[12] The average age was 37.4 years. Among all labor-years, 55 percent were male, and 22 percent had a child at home with the age under six-year olds. Among them, 8 percent were self-evaluated as in poor health conditions. In the survey, health condition is ranked in five categories: excellent, good, fair, poor, and disabled.

Compared to household migration, the individual migration rate was lower. Its average for 2004–2008 was 34 percent, whereas it was 63 percent for the household migration rate in the same period. This is nothing surprising; a migrant household usually had non-migrant labor at home. We also present the time trends of individual migration rates since 2004 in Figure 1. The major patterns are consistent with those of household migration rates. Specifically, the inexperienced migration rate remained flat; the experienced migration rate continued to grow, and comprised a larger and larger proportion of the total migrant labor force.

Table 5. Experienced migration propensities, 2004–2008.

Examination level:	Household level	Individual level						
Specification:	Baseline (1)	Baseline (2)	Age effect (3)	(4)	Gender effect (5)	(6)	Health effect (7)	All variables (8)
Age			-.0029***					
			(.00031)					
Age: 31–45				-.022***				-.018***
				(.0065)				(.0064)
Age: 46–60				-.089***				-.089***
				(.0097)				(.0095)
Being male					.028***	.017***		.027***
					(.0057)	(.0065)		(.0065)
Being male * Having a child						.05***		.047***
						(.014)		(.014)
Having a child						-.047***		-.051***
						(.012)		(.012)
In poor health condition							-.059***	-.044***
							(.015)	(.015)
Schooling	.0064***	.01***	.006***	.0062***	.0099***	.0099***	.0097***	.0054***
	(.0017)	(.0012)	(.0012)	(.0012)	(.0012)	(.0012)	(.0012)	(.0012)
Labor counts	.034***	-.0017	-.0057**	-.0033	-.00084	-.00018	-.002	-.00075
	(.0024)	(.0023)	(.0023)	(.0024)	(.0023)	(.0023)	(.0023)	(.0025)
Log (lagged landholding)	-.016***	-.007**	-.013***	-.012***	-.0074**	-.0068**	-.008***	-.013***
	(.0028)	(.0029)	(.003)	(.0029)	(.0029)	(.0029)	(.0029)	(.0029)
Log (lagged taxes)	-.0024*	-.0025*	-.0026**	-.0026*	-.0025*	-.0024*	-.0026**	-.0026*
	(.0013)	(.0013)	(.0013)	(.0013)	(.0013)	(.0013)	(.0013)	(.0013)
Log (lagged village fees)	.0016	.00081	.0013	.0013	.00076	.00066	.00062	.00097
	(.001)	(.001)	(.001)	(.001)	(.001)	(.001)	(.001)	(.001)
Lagged village migrant rates	.28***	.36***	.36***	.36***	.36***	.37***	.36***	.37***
	(.021)	(.027)	(.027)	(.027)	(.027)	(.027)	(.027)	(.027)
Year dummy: 2005	-.014*	-.0058	-.0045	-.0049	-.0057	-.0056	-.006	-.0049
	(.0083)	(.0087)	(.0086)	(.0086)	(.0087)	(.0087)	(.0086)	(.0086)
Year dummy: 2006	-.0081	.0088	.012	.011	.0089	.0086	.0087	.01
	(.0083)	(.0086)	(.0085)	(.0085)	(.0086)	(.0086)	(.0086)	(.0085)

Year dummy: 2007	-.018*	-.027***	-.023**	-.024**	-.027**	-.027**	-.028***	-.024**
	(.0099)	(.010)	(.010)	(.010)	(.010)	(.010)	(.010)	(.010)
Year dummy: 2008	-.019*	-.022**	-.016	-.018*	-.021**	-.021**	-.022**	-.017
	(.0098)	(.010)	(.010)	(.010)	(.010)	(.010)	(.010)	(.010)
Observations	14,562	20,682	20,682	20,682	20,682	20,682	20,682	20,682
Adjusted R^2	0.042	0.022	0.027	0.027	0.023	0.024	0.023	0.030

Note: Standard errors are in parentheses.

*$p < 0.1$; **$p < 0.05$; ***$p < 0.01$.

Our major focus is still on the instability of migrant labor supply. Accordingly, we report experienced laborers (those with the one-year lagged migration dummy equal to one) by their current migration status in Table 2. Compared to those that stopped migrating, laborers that continued migrating were more educated and had smaller land size at home. The difference in labor counts was minor. Regarding other individual characteristics, migrant laborers were younger, more likely to be male, and less likely to have children or to be in poor health conditions.

The summary statistics in Table 2 enable us to address two concerns about the sources of the instability of migrant labor supply. One possible explanation is that the instability was natural and mainly driven by retirement. Though it has been revealed that experienced laborers who stopped migrating were older, it is unlikely that retirement was the major reason; among these non-migrant laborers, the average age was only 35 years old, much younger than the retirement age. Additionally, the data show that 79 percent of laborers who stopped migrating were under the age of 46 years old, which can be hardly explained by retirement. The other possibility is that the major reason for stopping migrating was due to illness. As shown, only 6 percent of non-migrant experienced laborers were in poor health conditions. Therefore, illness was unlikely the major reason.

Table 3 presents individual migration propensities over provinces for 2004–2008. Summary statistics show that, conditional on having at least one migrant laborer, the average count of migrant laborers per household was 1.6 in 2004–2008. Consistently, we find that the average propensity of individual experienced migration was 6 percent lower than that of the household one, indicating that a substantial number of rural households sent out more than one migrant laborer. This difference was more pronounced in Hunan and Sichuan, two provinces that are well known for sending out enormous amounts of migrant laborers. We also present the time trends of experienced and inexperienced migration propensities at the individual level during the period 2004–2008 in Figure 2. It further confirms our impression that the propensity of experienced migration has stabilized since 2004.

Estimation results

We rerun the regression of experienced migration in 2004–2008 at the individual level in column (2) of Table 5. We find that both the household-level and the individual-level regressions lead to consistent estimation results, indicating that the concern that household-level regressions may bias the estimates is minor. Specifically, the individual-level regression shows a deeper drop of the propensity in 2007 and 2008. The education effect turns stronger as schooling is more precisely measured. Differently, the effect of labor counts is no longer positive; if there is any effect, it tends to be negative. The negative coefficient may reflect an effect of division of labor within a household: a household on average had three laborers, and when other family members were involved in migration, the examined laborer was more likely to stay to make up for the loss of rural labor at home.

The instability of migrant labor supply may be driven by the fluctuations of labor demand in the destination areas. Since we do not have information on destination areas, we cannot directly address this possibility. In this aspect, when we say the migrant labor supply is unstable, we regard it as an equilibrium phenomenon, which may be driven both by laborers' choice (actively quitting a job) and by negative demand shocks (being passively laid off). However, if the instability was mainly driven by negative demand shocks, it is less likely that the effects of individual characteristics were influential.

We thus turn to examine how individual characteristics were associated with the instability of migrant labor supply. We have ruled out the possibility that the instability was simply due to retirement. We further investigate how age influenced the propensity of experienced migration. Column (3) repeats the regression of column (2) with the inclusion of an additional variable, age of the laborer. The age effect is significantly negative, indicating that the instability was more severe among aged laborers.

To address the nonlinearity of this age effect, we replace the age variable with two age dummies and rerun the regression in column (4). One age dummy indicates that the laborer is at the age between 31 and 45, and the other indicates that he/she is at the age between 46 and 60, so the reference group is those laborers under 31 years old. As shown, the coefficients on both dummies are significantly negative, and their magnitudes are substantially different from each other. Compared to laborers under 31 years old, those at the age of 31–45 were 2.2 percent more likely to stop migrating. This effect was more influential among those at the age of 46–60. Their migration propensity was 8.9 percent lower than the youngest group, indicating that the instability was more severe among old migrant laborers.

One possible reason is the job-choosing-labor effect; that is, more stable job positions tend to be provided to younger laborers. As a result, older migrant laborers are more likely to hold part-time or less stable jobs. Anecdotic evidence has shown that original equipment manufacturer factories prefer hiring young workers, who are more able to do the repeated operations in the streamline. However, we tend to believe that the job-choosing-labor phenomenon, if it exists, should stem from the fact that the current institutional arrangement (here, the *hukou* system) somehow prevents migrant laborers from better accumulating their human capital. It is generally acknowledged that during economic recessions, junior laborers suffer from a higher unemployment rate (thus higher job instability) due to lack of relevant work experience compared to senior laborers. This argument apparently does not apply to the situation of China's migrant labor market.

Next, we examine the gender effect. Column (5) reruns the regression with the inclusion of a male dummy. The male dummy indicates whether the laborer is male; it equals one if so, and zero otherwise. As shown, its coefficient is significantly positive. Being male made a laborer 2.8 percent more likely to continue migrating. We further investigate whether this gender effect is related to the different gender roles in the family.[13] In a traditional rural household, the wife stays at home and does the housework while the husband cultivates in the field. Particularly, taking care of children is regarded as wives' duty. Column (6) thus includes two additional variables, a baby dummy and its interaction with the male dummy. The baby dummy indicates whether the household has at least one child under six-years old who needs intensive care; it equals one if so, zero otherwise. As shown, the coefficient on the baby dummy is significantly negative, indicating that having a baby at home resulted in a female laborer 5.0 percent less likely to continue migrating. In contrast, the coefficient on the interaction term is significantly positive, and its magnitude is comparable to that of the baby dummy, indicating that such an effect did not exist among male laborers. This confirms our expectation that wives take more responsibilities in raising children, thereby negatively influencing their migration persistence.[14] Additionally, the magnitude of the male effect drops by one-third. That is, the effect of having a child addressed about one-third of the instability difference between male and female laborers.

We have argued that illness should not be the major reason that migrant labor supply was unstable. We now examine how influential this health effect was. To do so, we repeat the regression by including a health dummy. The health dummy indicates whether the laborer is self-evaluated as in a poor health condition; it equals one if so, zero otherwise. As shown in column (7), the health effect is significant. A laborer with a poor health condition was 5.9 percent less likely to keep on migrating.

One may be concerned that estimating one effect at a time may fail to address the issue that these variables are correlated with each other. In column (8), we include all variables in one regression. As shown, the major results we have discussed remain unchanged. Overall, we find that younger laborers, male laborers, or laborers with better health conditions had higher migration persistence.

Concluding discussion

Using household survey data in rural China over the period 1986–2008, we analyze the instability of migrant labor supply from rural areas. To capture the circular nature of migration in China, we distinguish two types of migration – inexperienced migration and experienced migration, and focus on the latter. We first investigate migration dynamics at the household level by using the 87–99 subsample. We find that labor counts, education level, and network effects not only stimulated inexperienced house-holds to become migrant, but also improved their migration persistence. Our study also shows that the extent of migration persistence was relatively low, suggesting that migrant labor supply was unstable. Additionally, we find that migration instability was also attributed to local migration regulations.

We then estimate the probability of experienced migration both at the household level and at the individual level for the 04–08 subsample. We find that the pattern of migration instability was similar at both levels. The propensity of experienced migration has stabilized since 2004. Additionally, we find that migration instability was also attri-butable to individual characteristics, such as age, gender, and health conditions.

This paper contributes to the literature in three aspects. First, it is closely related to an important literature on migrant labor's mobility in the urban labor market, which is so far understudied. Using an urban household survey in 1999, Knight and Yueh (2004) investigated interfirm mobility in the urban labor market. They find that the mobility rate of migrant labor was significantly higher than that of urban labor. Our study provides additional evidence on this instability of migrant labor supply from source areas. We further investigate how it evolved over time and varied across regions, which may lead to a better understanding about the instability and its source.

Second, this paper is also related to the literature on determinants of circular migration in China. Topics in this field are typically focused on experienced laborers' decisions to permanently migrate (Hu, Xu, and Chen 2011), to return (Ma 2002; Wang and Fan 2006), and to re-migrate after having returned home (Wang and Zhao 2013). Different from these studies, we investigate circular migration from the angle of the instability of migrant labor supply. Our study also provides new insights on the mecha-nism behind labor shortages by documenting a potential labor supply crisis that happened in Guangdong Province in 1996. The instability of migrant labor supply is so far understudied. One exception is Knight and Yueh (2004). Using an urban household survey for 1999, Knight and Yueh (2004) find that Chinese migrant laborers had a very short job duration (2.2 years for completed employment), implying an annual turnover at the firm's aspect of 45.5 percent, which is higher than our estimate.

Third, this paper is also related to the studies on rural households' participation in the labor market. Based on a sample of 472 households in Zhejiang Province from 1995 to 2000, Brosig et al. (2007) apply a discrete-time hazard model to analyze rural households' status of labor market participation. Our paper differs from Brosig et al. (2007) by focusing on rural households' participation in the *urban* labor market with more comprehensive and recent data.

The perspective of the instability of migrant labor supply helps better understand the relocation of firms from coastal areas to inland areas, which has been happening in the recent years. A famous case is Foxconn, the world's largest contract manufacturer for major brand names such as Apple, Dell, and Toshiba. In 2010, Foxconn started to relocate its factories from Guangdong to inland areas. Three large production bases have been built in inland cities such as Chengdu, Zhengzhou, and Wuhan. Essentially, firms that choose to relocate are fighting against labor shortages, or to be more general, the instability of migrant labor supply.[15] Our story is that the *hukou* system resulted in the arrangement of migrant labor traveling back and forth, and that the associated commuting burden and lack of family time negatively influenced migration persistence. Therefore, it is reasonable to expect that, given everything else being equal (especially having the same stable jobs), experienced migration propensities should be higher when migrant workers are employed closer to their family.

We further provide some evidence by documenting regional differences of experienced migration propensities. Specifically, we compare Hunan with Guangdong: Hunan, an inland and less-developed province, sent a large number of migrant workers out to the neighboring province, Guangdong; while Guangdong, coastal and more developed, absorbed most of its own migrant workers. As expected, we find that in 2004–2008 the individual experienced migration propensity in Guangdong exceeded that in Hunan by 10 percent (90 percent vs. 80 percent as shown in Table 3). Of course, this difference may be because Hunan tended to provide unstable jobs to its migrant labor. It turns out this argument is partially right: among experienced laborers who had migrated out of Hunan in the previous year (932 out of 1531 experienced labor-years), the migration propensity was higher at the level of 84 percent. Even so, there was still a 6 percent gap to fill, compared to 90 percent – the propensity in Guangdong. As to those experienced laborers in Guangdong, few of them had migrated out of the province in the previous year (48 out of 2067 experienced labor-years). The gap between these two propensities increased to 6.8 percent (significant at the 1 percent level) when we control for all other factors mentioned in the paper. This gap should give us some idea about how much working close to the family would contribute to the stability of migrant labor supply with the presence of the *hukou* system.

Besides better reconciling the paradox of labor surplus and occasionally occurring labor shortages, our study may have other important implications. One that we have mentioned is its implication on migrant laborers' human capital accumulation. Another is on the Labor Contract Law enacted in June 2007 and in effect since January 2008. The purpose of the 2007 Labor Law may be to protect laborers from suffering from job instability. However, by ignoring that it is the institutional arrangement that results in such instability, the enforcement of the law may have a destructive effect, especially on rural-to-urban migration. With the presence of the *hukou* system, a reasonable labor contractual structure should be flexible enough to allow both sides to take the genuine instability into account. Simply requiring a strictly stated labor contract with less flexibility may put migrant laborers in a worse position when competing for jobs with urban laborers, thereby stopping them from migrating in the first place. The real effect needs to be evaluated when updated data become available.

Due to data limitations, so far we are silent about the instability of migrant labor supply in current China. To shed some light, we turn to the 2013 China Household Finance Survey (CHFS) data. The CHFS survey is a household finance-specific project first launched in 2011 based on a nationally representative sample. The project is directed by the Survey and Research Center for China Household Finance at the Southwestern University of Finance and Economics (SWUFE) in Chengdu. The survey provides information about household assets, income, expenditures, insurance, etc.

Though up to date, a major limitation of the 2013 CHFS data is that the survey did not ask about the number of migrant labor-days in the previous year. As a compromise, we use non-agricultural labor-days being greater than 180 days as a proxy for being migrant. This measure is less precise because a laborer who is mainly engaged in non-agricultural activities may stay at home rather than migrating. By overstating the denominator (e.g. the number of experienced laborers), the experienced migration propensity would be underestimated. Our calculation shows that the propensity of experienced migration was only 53 percent in 2013,[16] much lower than that in 2004–2008 based on the RCRE data. Because of the limitation, more reliable data are called for to precisely estimate the instability of migrant labor supply in current China.

Because the information about migration destination is limited, we measure the stability of migrant labor supply by the migration propensity of experienced laborers based on the observation from source areas. We thus regard a migrant labor as persistent as long as she migrates in two consecutive years. It may overstate the stability from the perspective of the destination firm, because a migrant laborer may migrate in two consecutive years but to different destination areas or to different firms in the same destination. In this aspect, our estimates can be regarded as an upper bound of the stability of migration labor supply.

Acknowledgments

We would like to thank Michael Greenstone, Steven Levitt, Robert Townsend, Katherine Miller, and participants at Empirical Microeconomics Workshop at University of Chicago for their helpful comments. We are indebted to Gale Johnson for his support and encouragement. We also thank Kam Wing Chan, Associate Editor, and one referee for their useful comments to greatly improve the manuscript. Data assistance from the National Bureau of Statistics of China is gratefully acknowledged.

Disclosure statement

No potential conflict of interest was reported by the authors.

Notes

1. Labor shortages seem to favor the argument that China has reached the Lewis turning point at which the modern economy has absorbed the surplus rural labor. However, the related debate is far from conclusive. Some researchers argue that China has reached the Lewis turning point based on evidence of rising migrant wages (Cai, Du, and Zhao 2007; Park, Cai, and Du 2007; Wang 2008). Others argue against it by pointing out that labor surplus in rural China still exists (Golley and Meng 2012; Kwan 2009; Minami and Ma 2009).
2. One exception is the literature on circulation (Newbold and Bell 2001). Chapman and Prothero (1983) study "circulation" in the Third World countries.

3. Practically, panel data are generally not available since it is extremely difficult to track and observe repeated migration decisions of the same person/household over time.

4. The pattern described in Figure 1 is different from that of Chan and Hu (2003). They describe the urbanization trend in from 1990 to 2000. With multiple measures, they depict that urbanization peaked in 1993–1995, dropped in 1996–1998, and has picked up since 1999 (figure 3a, page 60). The pattern of inexperienced migrants is similar, with a dip in 1997–1998, and the patterns of experienced and all migrants are distinct. There might be a couple of reasons for the discrepancy. First, our data are missing in the years of 1992 and 1994. Due to the dynamic model specification, observations in years of 1992–1995 are dropped. We refrain from describing any pattern for the 1992–1995 interval. Second, our sample may not coincide with the sample in Chan and Hu (2003). One of the measures they use is the number of rural workers in urban work-units. Our survey investigated the migrants from the perspective of source areas; as long as a household sends people out for non-farm activities, they are categorized as migrant.

5. The formalized version of the *hukou* institution can be traced back to a far-reaching *hukou* regulation (*hukou dengji tiaoli*) in 1958. The regulation remains fully in force even today (Zhu 2003). It requires that any migration is subject to approval from the destination authorities. For the evolution of China's *hukou* system, please refer to Chan (2009) and Chan and Buckingham (2008).

6. This model does not include leisure and is silent about the trade-off between income and substitution effects. We choose to abstract from leisure for two reasons. First, there is no good measure of quantity and quality of leisure. Second, it makes sense to assume that the substitution effect dominates the income effect for a representative rural household in a low-income country such as China.

7. Admittedly, this classification is somewhat arbitrary. As a robustness check, we repeat the regressions using a lower threshold (150 labor-days). The major results persist.

8. As a robustness check, we repeat the regressions with the inclusion of outliers. The major results persist.

9. We do not claim the causality here. It is well known that it is hard to identify the causality when local spillovers are examined. A number of authors have noted the importance of chain migration patterns in China. Rozelle, Tayor, and deBrauw (1999) find that an established network in the destination leads to new migration of the same magnitude as the existing network size. Giles (2006) finds that having access to village migrant networks significantly increases household incomes and improves the ability of rural households to smooth consumption.

10. According to the 1995 census (1 percent sample), migrants from Hunan and Sichuan accounted for the largest interprovincial migrant population to Guangdong. Among those 18,859 people who were originally out of Guangdong but later obtained residency in Guangdong, 4468 (23.7 percent) and 3645 (19.3 percent) were from Hunan and Sichuan, respectively.

11. This is how the NBS defines rural laborers. The same standard is followed when we calculate labor counts and average schooling at the household level since 2003. As a robustness check, we also define laborers as those who are not disabled, not students, and with the age of 16–59. We rerun the regressions and the major results persist.

12. See Appendix 1 for the details of the data screening process. The sample size looks smaller when it is compared with the product of the sample size at the household-level and average labor counts. The reason is that an observation is valid only if it has a match in the previous year. It is less likely to have such a match at the individual level than at the household level.

13. Again, one may attribute it to job market discrimination against women.

14. Using the 2008 RUMiC Rural Household Survey data, Lee and Meng (2010) examine the determinants of rural-to-urban migration. They find that having one more child aged 0–4 years results in a drop of 5 percentage points on women's migration probability. They interpret it as evidence of how China's institutional restrictions influence migration decisions. Due to the difficulty in obtaining access to social services and welfare, migrant laborers generally leave their family behind. Women, who are supposed to have more family responsibilities, are thus less likely to migrate.

15. It has been argued that this relocation is driven by the rising labor wage in coastal cities. However, the wage savings may not cover the increasing cost of shipping due to the greater distance from harbors.

16. To begin with, we restrict the sample to rural households, with the exclusion of those who were not in the labor market in 2013 (that is, student, housewife, disable, retired, and those younger than 16 or older than 80). We end up with a sizable sample of 22,076 rural-*hukou* laborers. A laborer is defined as migrant if she was living in an urban area in 2013. A migrant labor is defined as inexperienced if she did farm work for more than 180 labor-days in 2012; otherwise, she is defined as experienced.

References

Benjamin, Dwayne, Loren Brandt, and John Giles. 2005. "The Evolution of Income Inequality in Rural China." *Economic Development and Cultural Change* 53: 769–824.

Borjas, George J. 1999. "The Economic Analysis of Immigration." In *Handbook of Labor Economics*, Vol. 3, edited by Orley Ashenfelter and David E. Card, 1697–1760. Amsterdam: North-Holland.

Brosig, Stephan, Thomas Glauben, Thomas Herzfeld, Scott Rozelle, and Xiaobing Wang. 2007. "The Dynamics of Chinese Rural Households' Participation in Labor Markets." *Agricultural Economics* 37: 167–178.

Cai, Fang, Kam Wing Chan. 2009. "The Global Economic Crisis and Unemployment in China." *Eurasian Geography and Economics* 50: 513–531

Cai, Fang, Yang Du, and Changbao Zhao. 2007. "Regional Labor Market Integration since China's WTO Entry: Evidence from Household-Level Data." In *China: Linking Markets for Growth*, edited by R. Garnaut, and Ligang Song, 133–150. Canberra: Asia Pacific Press.

Cai, Fang, Albert Park, and Yaohui Zhao. 2008. "The Chinese Labour Market in the Reform Era." In *China's Great Economic Transformation: Origins, Mechanism, and Consequences*, edited by L. Brandt and T. G. Rawski, 167–214. Cambridge: Cambridge University Press.

Chan, Kam Wing. 2001. "Recent Migration in China: Patterns, Trends, and Policies." *Asian Perspectives* 25: 127–155.

Chan, Kam Wing. 2009. "The Chinese Hukou System at 50." *Eurasian Geography and Economics* 50: 197–221.

Chan, Kam Wing. 2010. "A China Paradox Migrant Labor Shortage amidst Rural Labor Supply Abundance." *Eurasian Geography and Economics* 51: 513–530.

Chan, Kam Wing. 2012. "Migration and Development in China: Trends, Geography and Current Issues." *Migration and Development* 1 (2): 187–205.

Chan, Kam Wing. 2013. "China Internal Migration." In *The Encyclopedia of Global Human Migration*, Vol. II, edited by Immanuel Ness and Peter Bellwood, 980–995. Boston: Blackwell.

Chan, Kam Wing, and Will Buckingham. 2008. "Is China Abolishing the Hukou System." *China Quarterly* 195: 582–606.

Chan, Kam Wing, and Ying Hu. 2003. "Urbanization in China in the 1990s: New Definition, Different Series, and Revised Trends." *China Review* 3: 49–71.

Chang, Hongqin, Xiao-yuan Dong, and Fiona MacPhail. 2011. "Labor Migration and Time Use Patterns of the Left-behind Children and Elderly in Rural China." *World Development* 39: 2199–2210.

Chapman, Murray, and R. M. Prothero. 1983. "Themes on Circulation in the Third World." *International Migration Review* 17: 597–632.

Chen, Guifu, and Shigeyuki Hamori. 2009. "Solution to the Dilemma of the Migrant Labor Shortage and the Rural Labor Surplus in China." *China & World Economy* 17: 53–71.

Fan, C. Cindy. 2009. "Flexible Work, Flexible Household: Labour Migration and Rural Families in China." *Research in the Sociology of Work* 19: 377–408.

Farre, Lidia, and Francesco Fasani. 2011. "Media Exposure and Internal Migration – Evidence from Indonesia." *IZA working paper*.

Giles, John. 2006. "Is Life More Risky in the Open? Household Risk-coping and the Opening of China's Labor Markets." *Journal of Development Economics* 81: 25–60.

Golley, Jane, and Xin Meng. 2012. "Has China Run out of Surplus Labor." *China Economic Review* 22: 555–572.

Hsiao, Cheng. 1986. *Analysis of Panel Data*. Cambridge: Cambridge University Press.

Hu, Feng, Zhaoyuan Xu, and Yuyu Chen. 2011. "Circular Migration, or Permanent Stay? Evidence from China's Rural–Urban Migration." *China Economic Review* 22: 64–74.

Knight, John, Quheng Deng, and Shi Li. 2011. "The Puzzle of Migrant Labour Shortage and Rural Labour Surplus in China." *China Economic Review* 22: 585–600.

Knight, John, and Linda Yueh. 2004. "Job Mobility of Residents and Migrants in Urban China." *Journal of Comparative Economics* 32: 637–660.

Kwan, Fung. 2009. "Agricultural Labour and the Incidence of Surplus Labour: Experience from China during Reform." *Journal of Chinese Economic and Business Studies* 7: 341–361.

Lee, Leng, and Xin Meng. 2010. "Why Don't More Chinese Migrate from the Countryside? Institutional Constraints and the Migration Decision." In *The Great Migration: Rural–Urban Migration in China and Indonesia*, edited by X. Meng, C. Manning, S. Li, and T. Effendi, 23–46. Cheltenham: Edward Elgar Publishing Ltd.

Liang, Zai, and Zhongdong Ma. 2004. "China's Floating Population: New Evidence from the 2000 Census." *Population and Development Review* 30: 467–488.

Ma, Zhongdong. 2002. "Social-Capital Mobilization and Income Returns to Entrepreneurship: The Case of Return Migration in Rural China." *Environment and Planning A* 34: 1762–1784.

Massey, Douglas S., Joaquin Arango, Graeme Hugo, Ali Kouaouci, Adela Pellegrino, and J. Taylor. 1993. "Theories of International Migration: A Review and Appraisal." *Population and Development Review* 19: 431–466.

Minami, Ryoshin, and Xinxin Ma. 2009. "The Turning Point of Chinese Economy: Compared with Japanese Experience." Conference paper, Tokyo, ADBI.

National Bureau of Statistics. 2014. *Monitoring Survey Report of Rural Migrants*. www.stats.gov.cn.

Newbold, K. Bruce, and Martin Bell. 2001. "Return and Onwards Migration in Canada and Australia: Evidence from Fixed Interval Data." *International Migration Review* 35: 1157–1184.

Park, Albert, Fang Cai, and Yang Du. 2007. "Can China Meet Her Employment Challenges?" Conference paper, Stanford: Stanford University.

Rozelle, Scott, J. Taylor, and Alan deBrauw. 1999. "Migration, Remittances, and Agricultural Productivity in China." *American Economic Review* 89: 287–291.

Todaro, Michael P. 1969. "A Model of Labor Migration and Urban Unemployment in Less Developed Countries." *American Economic Review* 59: 138–148.

Todaro, Michael P. 1976. *Internal Migration in Developing Countries: A Review of Theory, Evidence, Methodology and Research Priorities*. Geneva: International Labor Office.

Wang, Dewen. 2008. "Lewisian Turning Point: Chinese Experience." In *Reports on Chinese Population and Growth No. 9: Linking up Lewis and Kuznets Turning Points*, edited by Fang Cai, 81–103. Beijing: Social Sciences Academic Press (in Chinese).

Wang, Wenfei, and Cindy Fan. 2006. "Success or Failure: Selectivity and Reasons of Return Migration in Sichuan and Anhui, China." *Environment and Planning A* 38: 939–958.

Wang, Zicheng, and Zhong Zhao. 2013. "Nongmingong qianyi moshi de dongtai xuanze." [Dynamic Pattern Choices of Migrant Workers.]) *Guanli Shijie* [Management World] 1: 78–88.

Wang, Feng, and Xuejin Zuo. 1999. "Inside China's Cities: Institutional Barriers and Opportunities for Urban Migrants." *American Economic Review* 89: 276–280.

Zhao, Yaohui. 1999. "Leaving the Countryside: Rural-to-Urban Migration Decisions in China." *American Economic Review* 89: 281–286.

Zhu, Lijiang. 2003. "The Hukou System of the People's Republic of China: A Critical Appraisal under International Standards of Internal Movement and Residence." *Chinese Journal of International Law* 2: 519–565.

Appendix 1. Data screening process

Table A1 shows how the original data are screened step by step, both at the household level and at the individual level. The first line of Table A1 presents the sample size of the original data. The bottom line corresponds to the sample size in Table 1.

Table A1. Data screening process.

	Household sample			Individual sample
	Full sample	87–99 (6 provinces)	04–08 (11 provinces)	04–08 (11 provinces)
Before screening	72,420	38,365	34,055	139,602
Step 0: Dropping non-labor obs.	N/A	N/A	N/A	(−)46,500
Step 1: Lag specification	(−)12,251	(−)6325	(−)5926	(−)16,165
Step 2: Dropping labor counts = 0	(−)2799	(−)264	(−)2535	N/A
Step 3: Dropping invalid obs.	(−)1965	(−)1022	(−)943	(−)11,609
Step 4: Keeping 0.5–99.5 percent	(−)1162	(−)520	(−)642	(−)1035
After screening	54,243	30,234	24,009	64,293

Appendix 2. Data comparison

We first examine the representativeness of the 03–08 data at the household level. To do so, we compare the 03–08 data of 11 provinces with the nationwide RCRE survey data. Our data are part of the nationwide RCRE data. Table A2 compares the means of three variables, household size, labor counts, and landholding in each year. To avoid the influence of extreme values, in our calculation we set the value invalid if the variable is either less than the 0.5 percentile or greater than the 99.5 percentile. As shown, the two demographic characteristics, household size and labor counts, are very close between these two data. The land size of our sample is somehow higher than the full sample in the first three years, but lower in the last three years. Overall, our sample of 11 provinces is not very different from the full sample of the survey.

Table A2. Household-level comparison with the national RCRE data.

Year	Household size		Labor count		Landholding	
	Our data	National	Our data	National	Our data	National
2003	4.09	4.11	2.73	2.82	5.74	5.69
2004	4.11	4.12	2.76	2.89	6.12	5.74
2005	4.10	4.13	2.74	2.90	5.99	5.77
2006	4.10	4.12	2.73	2.89	5.81	5.95
2007	4.09	4.10	2.71	2.90	5.41	6.09
2008	4.07	4.10	2.68	2.88	5.08	5.90

Note: The summary statistics of the nationwide RCRE data come from the book, *Statistics of the National Rural Survey on Fixed Observation Points (2000–2009)* (*Quanguo Nongcun Guding Guancedian Diaocha Shuju Huibian*).

We then examine the representativeness of our data at the individual level by comparing our labor sample with the labor sample from the 2005 census (1 percent sample) data. The 1 percent 2005 census data have a total of 1363,047 rural residents, while our data have 23,351 in 2005. We examine the age distribution and the education distribution as follows.

We first investigate the age distribution. Figure A1 presents the comparison of age distributions among laborers. In the 1 percent 2005 census data, there are 841,760 laborers; in our data, there are 16,215 laborers in 2005. Consistent with the definition of rural laborers by the RCRE survey, we define laborers as males of 16–60 years old and females of 16–55 years old. As shown, the patterns of the two distributions are similar. The dip around 45 years old can be traced back to the 1959–1961 famine. The sharp drop in 56 is because women over 55 years old are no longer counted as laborers. Figure A1 also shows that our sample tends to have a greater number of younger laborers compared to the 1 percent 2005 census data. This difference may be due to different sample coverage: the RCRE survey is household-based and includes all members in the household, while the census is based on individuals' residency (typically, an individual is included when the individual lives in the area for more than six months). Younger rural laborers were more likely to migrate, thus were less likely to be counted as rural residents in their source areas by the census.

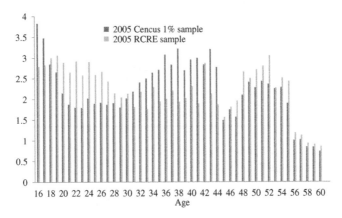

Figure A1. Age distributions of labor in 2005.

We then compare the distribution of education in these two labor samples. In the 1 percent 2005 census data, the valid sample size is still 841,760; in our data, there are 15,672 laborers in 2005 with the education variable valid. Instead of reporting the years of education, the census data categorize the level of education into seven groups. The proportions of the highest three groups (associate, college, and post-graduate) are small, and thus we treat them as one group. To be comparable, we fit the number of education years into the five categories.

As shown in Table A3, the proportions in each category are close between these two samples. However, the proportion of illiterate is higher and the proportion of senior high is lower in the census data. This is consistent with the finding that our sample included a greater number of younger individuals, and younger individuals tend to be better educated.

Table A3. Education distributions of labor in two samples.

Education level	2005 census data		Our 2005 data	
	Observations	Proportion (percent)	Observations	Proportion (percent)
Illiterate (0–1)	67,233	7.99	742	4.73
Elementary (2–6)	278,708	33.11	4941	31.53
Junior (7–9)	410,403	48.76	7518	47.97
Senior (10–12)	75,819	9.01	1959	12.51
College or above (>12)	9597	1.14	512	3.27
Total	841,760	100	15,672	100

Residential mobility within Guangzhou city, China, 1990–2010: local residents versus migrants

Si-ming Li[a] and Yushu Zhu[b]

[a]Department of Geography, Hong Kong Baptist University, Kowloon Tong, Hong Kong; [b]Population Studies and Training Center, Brown University, Providence, RI, USA

Drawing on residential history data from two household surveys conducted in Guangzhou in 2005 and 2010, this paper compares the pattern of intra-city residential moves of local residents and that of migrants. The findings show different trajectories of residential moves for the two groups. While migrants showed increasing mobility over time, residential moves of locals first rose until the early 2000s, then declined steadily afterward. Moreover, the determinants of residential moves of migrants differ from those of the local population. Also, whereas residential moves for the local population are subject to changing factors over time, drivers of relocation for migrants remain more or less stable.

Introduction

Residential mobility within a city constitutes a major component of urban spatial dynamics and is instrumental to the continuity and change of urban neighborhoods. At the personal level, prolonged residence in the same dwelling or neighborhood is a key to developing deep affection to a place. But, the inability to relocate could also mean the difficulty in adjusting to changing circumstances such as neighborhood decline and en masse relocation of job opportunities in the urban area, hence the spatial mismatch hypothesis (Kain 1968, 1992). For socially deprived groups, the inability to move out of segregated neighborhoods despite their frequent moves is symptomatic of and feeds upon the culture of poverty (Rosenbaum, Reynolds, and Deluca 2002; Wilson 1987).

China's market-oriented reforms over the past decades has major implications for the way urban housing is supplied and consumed and hence, the distribution and redistribution of population over the urban space. The process of inter-city migration and the underlying mechanisms pertaining to household registration (*hukou*) system have been well documented (e.g. Chan, Liu, and Yang 1999; Chan and Zhang 1999); the literature, however, has paid much less attention to the more local process of intra-city migration. Indeed, only a few authors have analyzed the patterns of and factors underlying housing decisions and residential mobility of China's urbanites (see e.g. Huang and Deng 2006; Li 2004, 2005; Li and Siu 2001; Li, Wang, and Law 2005; Logan, Bian, and Bian 1999; F. Wu 2004). The focus of this literature is on people with the proper residential status, or local *hukou*, in the city. Much less is known about the extent to which the hundreds of millions of non-*hukou* migrants[1] (Chan, Liu, and Yang 1999) from China's vast rural hinterlands change residence and under what conditions they relocate after having arrived in a major city. Do migrants move in response to the realignment of job and housing opportunities in the city under an increasingly neoliberal labor and housing market regime (He and Wu 2009)? Or, are the majority forced to move due to eviction by landlords, exorbitant rents, and redevelopment of low-cost inner-city neighborhoods and of villages-in-the-city (VICs) where migrants congregate?

Studies conducted in the West reveal that immigrants tend to have relatively high mobility rates when they first settle in the place of destination, but the propensity to move decreases subsequently and finally approaches the level prevailing in the host society (Owusu 1999; Renaud and Bégin 2006). The theory of residential assimilation further postulates that in due course, maybe over one or two generations, migrants and their descendants will move out of segregated ethnic communities to join the ranks of the host society (Alba and Nee 1997). However, the migration models pertaining to market economies in the West may have limited applicability in China due to different socio-political contexts (Chan, Liu, and Yang 1999). More specifically, the household registration (*hukou*) system with its attendant social policies has been deployed by the state to constrain rural-to-urban migration (Cai and Chan 2009), and thus has served as a persistent institutional barrier for migrants' residential assimilation. As such, residential behavior and outcomes may show rather different patterns between migrants and local residents.

We make use of retrospective residential history data gathered from two large-scale surveys conducted in the City of Guangzhou in 2005 and 2010 to reveal the patterns of intra-city moves over the period 1990–2010, juxtaposing the moves of migrants against permanent residents (also referred to as local population in the study). Guiding the analysis is the event–history approach, which arguably provides a better account of the triggers behind residential moves than analyzing cross-section snapshots (Clark and Dieleman 1996). Below, we first provide a brief review of China's urban housing reform. This helps to delineate the context under which housing consumption and residential relocation were undertaken in different phases of the reform. Then, we present the conceptual framework, drawing on the existing literature on residential behavior. This helps identify the variables used in the discrete-time logit analysis – the modeling approach adopted for this study. Next, we describe the data-set and provide descriptive statistics of the variables employed. The main corpus of analytical findings is given in two sections. The first section charts the change in residential mobility rates over the study period for migrants and local population of the city. The second section reports the results of the statistical analysis on the determinants of residential mobility.

The research context

Housing reform in urban China

While a number of reform experiments were undertaken in the early 1980s, until the early 1990s housing in urban China was mainly distributed through state work units (*danwei*) as remuneration in kind; the private housing market barely existed. Nonetheless, professional developing companies had already replaced the individual work units as the main agents in new housing provision (Wang and Murie 1996; Wu 1996).

The system of socialist provision of urban housing was gradually dismantled in the 1990s. The "Decision on Deepening the Urban Housing Reform" promulgated in 1994 called for the establishment of a two-tier housing provision system: affordable housing (*jingjishiyong fang*) for low- and middle-income households and commodity housing (*shangpin fang*) for higher-income groups. A mandatory Housing Provident Fund was instituted that first applied to workers of state work units and then also to private and foreign enterprises. At the same time, state-owned commercial banks were authorized to extend mortgage loans to assist home purchases (Li and Yi 2007). A functioning commodity housing market began to take shape.

The welfare provision of housing formally came to an end with the issuing of "On Deepening Urban Housing System Reform and Quickening Housing Construction" by the State Council in 1998. The turn of the twenty-first century saw wholesale transfer of work-unit housing to their workers through heavily discounted sales, which served to perpetuate housing inequalities inherent in the former housing allocation system. Conferment of full property rights and hence windfall financial gains were given to owners of privatized work-unit housing, variously known as reformed housing, in the late 1990s and early 2000s.

The past decade witnessed heated housing speculation, housing price inflation, and massive commodity housing developments (Hui and Yue 2006). Municipal governments' monopoly over the primary land market; i.e. having the sole authority given to them by the central government to requisition land from rural villages and convert the land to urban usage, has enabled them to reap huge rent gaps or the differences between the value of the land under agricultural and urban use. Land has become the single-most important source of revenue for municipal governments through property-led urban development (Fu and Lin 2013; He and Wu 2007; Zhu 1999).

Rising home prices have caused widespread discontent. Unlike their parents who were able to achieve ownership through the purchase of rental public housing at huge discounts, the younger generations today find homeownership an increasingly difficult proposition without parental support (Li 2010; Li and Yi 2007). Also, rampant property speculation presents a real risk of a property bubble burst and the implied devastating consequences to the national economy. In recent years the central government has taken repeated measures to dampen housing property speculations, such as higher mortgage interest rates charged to buyers of second and subsequent homes and restricting home sales only to individuals with the local *hukou* (Li and Du 2014). Furthermore, after decades of market rhetoric the central government is once again stressing the need for social or public housing. Extensive financial incentives have been given to local governments to construct low-rent housing (*lianzu fang*) and expand the affordable housing programs (Wang and Murie 2011). However, these policies primarily apply to residents with the local *hukou*.

Housing for migrants in Chinese cities

Growing rural-urban and inland-coastal disparities have caused hundreds of millions of peasants in China's vast rural hinterlands in lagging regions to migrate to the coastal metropolises in search of jobs and better futures. It is estimated that the number of migrant workers with the rural *hukou* jumped from 104 million in 2002 (Cai and Chan 2009) to 150 million in 2010 (National Bureau of Statistics of China 2010). By the early 2000s, rural migrants already accounted for 30 percent of the total urban labor force (Cai and Chan 2009). It is well documented that migrants, rural migrants in particular, have encountered major difficulties in accessing housing in cities (Solinger 1993; Wang and Murie 2000; Wu 2006). In the past, the lack of the local *hukou* excluded them from socialist welfare housing (Wang and Murie 2000; Wu 2006). Today, in theory, migrants could rent or purchase commodity housing; however, meager incomes and precarious employment largely preclude them from accessing decent housing in the market (Zhu 2014; Zhu, Fu, and Ren 2014). A substantial proportion of migrant workers live in factory dormitories and make-shift structures in construction sites (Wang 2000; Wang and Zuo 1999). Others seek informal housing in VICs.[2] VICs first appeared in the late 1980s and early 1990s on the peripheries of fast-growing metropolises such as Beijing and Guangzhou. Housing in VICs is usually of sub-standard quality and subject to fire hazards and poor hygiene conditions. Yet, they have become the single-most important housing source for rural in-migrants (Wang, Du, and Li 2014). The development of VICs reached its peak in the late 1990s. By 2000, in Guangzhou there were 277 VICs accommodating some one million dwellers, the great majority being rural in-migrants. In Shenzhen, the corresponding figures were 241 VICs and over two million inhabitants (Song, Denou, and Ding 2008).

In recent years, massive (re) development in conjunction with place-making efforts and real estate speculation have eradicated large numbers of VICs in the city center (He and Wu 2007). For example, in Beijing 171 VICs within the Fourth Ring Road and those adjacent to Olympics stadium sites were torn down to make way for the 2008 Olympics, and another 61 VICs were to be cleared in the next two years. In Guangzhou, the municipal government issued a VIC redevelopment plan in 2009, which identified 52 VICs to be redeveloped within the next three to five years. Large-scale disappearance of VICs in China's metropolises will certainly intensify the housing problem for migrants, especially those from rural areas (Zhu 2014).

Literature review and the conceptual framework

Determinants of intra-city residential moves

Ever since the publication of Rossi's (1955) seminal piece, *Why Families Move*, residential mobility has been primarily seen as a spatial adjustment process through which individual households change residence in order to meet changing housing needs and preferences. Scholarly discussion of the determinants of intra-city residential moves has focused on two themes: life cycle and economic rationality. Life-cycle or life-course events such as marriage, arrival and departure of children in the family, retirement, and age-related health matters have been identified as major causes behind residential moves. Age is reported to have a curvilinear relationship with mobility. Effects of the current marital status are ambiguous, although change in marital status increases household mobility. Household size does not show clear effects on mobility, but having school-aged children tends to decrease the propensity to move.

The model of economic rationality regards residential moves as an attempt to restore housing consumption equilibrium, juxtaposing between moving costs and benefits. Household characteristics correlated with this model usually include socioeconomic status (income and education) and employment status (occupation, job change, promotion). Income growth in association with advancement along the career path, job change, and the availability of the private car also affect residential and locational preferences and result in a house move. Housing tenure is closely related to the propensity to move, as the cost of moving under rental occupation is much lower than under ownership. Young adults with unstable jobs or having unstable partnership tend to choose renting in anticipation of frequent moves; on the other hand, family households with children in school are more inclined to owner occupation and staying in the same house in a suburban neighborhood for an extended period of time.

Residential mobility or the lack of it not only reflects individual choice outcomes, but is also indicative of the interplay of more macrostructural and institutional forces. Examples include the pervasive and persistent racial discrimination in the United States, the prevalence of social housing in many European countries, the residualization of council housing in the United Kingdom under the "Right-to-Buy" policy, and the neo-liberal reforms, which have tremendously heightened job insecurity as well as social and spatial inequality in most market economies since the 1980s. In China, given its entrenched socialist planned economy legacies, structural and institutional forces continue to play an important part in delineating residential choices and relocation possibilities. These include, *inter alia*, segmented housing markets, the dual household registration system, and dual labor markets, which would distort the demand-oriented predictions of relocation patterns for Chinese households. For instance, contrasting the norm in more mature market economies, Huang and Deng (2006) reveal that owners were more mobile than renters in the early reform period, as most urban residents stayed in rental housing provided by the work units, which would discourage residential mobility. Also, Li (2004) and Li, Wang, and Law (2005) reported that affiliation to state organizations and state-owned enterprises, and membership in the Chinese Communist Party (CCP), which facilitated moving up the housing ladder, were associated with higher mobility rate. With market forces assuming increasing importance, life-cycle factors as well as income and wealth are now playing an increasing role in the residential behavior of Chinese urbanites (Huang and Deng 2006; Li 2004).

For (rural) migrants, the institutional barriers embedded in the dual *hukou* system are even more formidable (W. Wu 2004). Former municipal or work-unit housing could only be acquired by sitting tenants who had work-unit affiliations and were local urban residents; likewise, economic or affordable housing is reserved for those with the local *hukou*. The segmented housing market and institutional marginalization with respect to housing access as previously reviewed would consequently affect the relocation behavior of migrants. Whereas in more recent years, residents with the local or urban *hukou* have been observed to move in response to housing demands, migrants tend to remain passive agents in the urban housing market. Wu (2006) observed that migrants do not move to pursue home ownership or housing preferences; socioeconomic factors could only account for a small fraction of their intra-city moves. For them, proximity to job or work-related events are found to be the strongest driving forces behind residential moves. Due to their transitory position in the city, lack of ownership, and unstable employment status, migrants are expected to move more frequently than permanent residents; yet, at the same time they tend to be spatially trapped in run-down neighborhoods, VICs, shanty towns (*penghuqu*), and dilapidated former work-unit compounds (Wang and Zuo 1999; W. Wu 2004).

The conceptual framework

Given the historical context of housing reform and the dual *hukou* system, we approach internal residential mobility in Chinese cities from two dimensions. First is a temporal dimension. We argue that intra-city migratory patterns for urban residents are subject to period effects. That is, we expect to find temporal changes in the patterns and determining factors of residential moves for the general population over different stages of the housing reform; more specifically, market-related factors would become more visible in more recent years, especially for local residents. Second is a comparative dimension. We believe that housing disparities between locals and migrants would lead to quite distinct rationales behind moving decisions. In particular, for local residents both institutional affiliation (e.g. work unit) and housing needs and preferences would assume importance; for migrants, housing preferences or needs would have minimal effects on the decision to move.

Data

This study mainly draws on data from two household surveys conducted in Guangzhou in 2005 and 2010. Both surveys adopted essentially the same multi-level probability-proportional-to-size sampling strategy to ensure the spatial representativeness of the data. The 2005 survey covered the original eight urban districts of Guangzhou plus the northern part of Panyu District; whereas, the 2010 survey extended the geographical coverage to Qiaonan Sub-District in south Panyu in line with urban expansion.

The two surveys employed broadly the same questionnaire, thus enabling the pooling of data from the two data-sets. Among the data collected was retrospective information on residential and employment histories of the household head. In the 2005 survey, the recalls dated back to 1980; in the 2010 survey, they dated back to 1990. A caveat has to be noted in the use of retrospective data – in addition to recall errors, retrospective data are subject to cohort effects, as cohorts of earlier years might have left the city or died and were therefore not included in the sample (Li 2004). As such, estimates of residential mobility rate based on retrospective life-history data tend to be biased downward. Recall errors may be less serious in the study of residential history in China because of the low residential mobility rates before the late 1990s (Huang and Deng 2006; Li 2004). To minimize recall errors, in the subsequent analysis we only examine intra-city residential moves after 1990 and for a maximum of three moves.

Based on the location of *hukou* registration, the survey respondents can be divided into *locals* and *migrants*. A *local* refers to one with the Guangzhou *hukou* and a *migrant* refers to one without it. Note that both samples are targeted at people residing in permanent residences and therefore exclude migrants residing in dormitories and construction sites who are probably more mobile than others.

The 2005 survey comprises 1203 households. Local *hukou* accounts for 90.2 percent, whereas non-local *hukou* 9.8 percent. Within the latter category, 73.5 percent are migrants with the agricultural *hukou*. For the 2010 sample, of the 1250 households, the share of local *hukou* and non-local *hukou* are 64.7 and 35.3 percent, respectively (Table 1). Among the latter, approximately one-third are urban migrants and two-third are rural migrants (Table 2). Recall that migrants living in factory dormitories are

Table 1. Distribution of locals vs. migrants in Guangzhou city.

	Locals	Migrants
Total sample	1893 (77.2%)	558 (22.8%)
2005 main sample	1084 (90.3%)	117 (9.7%)
2005 1% national survey	76.3%	33.7%
2010 main sample	809 (64.7%)	441 (35.3%)
2010 census	64.0%	36.0%

Source: Survey data; tabulation on 2005 1percent National Population Sample Survey; manual of Guangzhou statistics information 2013 (Guangzhou Statistical Bureau 2013).

excluded in both surveys. Perhaps indicative of migrants' increasing reluctance to reside in factory dormitories as well as employers' concerns about managing the dormitories, the 2010 sample is more in line with the share of migrants (36 percent) in the total city population (Guangzhou Statistical Bureau 2011). To address the problem of possible under representation of migrants in the surveys employing official records of distribution of households over geographical districts as the sampling frame, in 2005 and 2010 surveys, a supplementary sample of 300 migrant households was drawn from 12 VICs, where migrants were concentrated.

All respondents in the surveys are adults over 19 years old (students constitute less than 0.2 percent in both instances). Tabulations based on the 2005 and 2010 main samples (Table 2) show that migrants from both urban and rural areas are much younger (more than 70 percent below 40 years old in both samples) than are locals (about 40 percent in the 2005 sample and 60 percent in the 2010 sample are older than 40 years of age). Also, whereas more than 80 percent of locals own a home in both samples, the rates of home ownership for urban migrants as given by the 2005 and 2010 samples are 41.9 and 30.2 percent, and that for rural migrants are 17.4 and 13.4 percent, respectively. Between the two, the higher homeownership rate for urban migrants can be attributed to their much higher levels of educational attainment, as compared with those of rural migrants. In fact, in terms of educational attainment, urban migrants are quite

Table 2. Composition of the main sample.

	2005 main sample				2010 main sample			
	Urban local	Rural local	Urban mig.	Rural mig.	Urban local	Rural local	Urban mig.	Rural mig.
No. of obs.	1053	31	31	86	785	24	149	292
Age								
≤20	4	–	2	–	2	–	2	9
21–30	260	9	10	41	88	4	67	112
31–40	380	13	11	35	220	3	49	100
41–50	306	5	6	9	231	7	18	50
51–60	87	1	2	1	135	4	7	15
>60	16	3	–	–	109	6	6	6
Education								
Illiterate	3	1		–	16	1	3	6
Primary school	48	1		1	59	6	2	44
Junior secondary	185	10	4	35	184	9	29	119
Senior secondary	489	12	13	41	235	5	53	88
Tertiary degree	175	5	7	6	95	1	24	16
College	130	2	4	2	167	2	35	19
Above college	23	–	3	1	29	–	3	–
Marital status								
Never married	216	7	12	16	73	5	37	66
Ever married	837	24	19	70	712	19	112	226
Homeownership								
Owner	863	23	13	15	641	21	45	39
Other	190	8	18	71	142	3	102	250

Source: Computations based on survey data.

similar to locals, if not higher than the latter. In the 2010 sample, 41.6 percent of urban migrants received post-secondary or above education; the corresponding figure for locals is 37 percent. Yet, urban migrants' homeownership rate remains some 50 percentage points lower than that of locals. Apparently, socioeconomic status alone cannot explain the bulk of variations in the homeownership rate, the latter being a major covariate of residential mobility. *Hukou* continues to be of major importance.

The age, education, and homeownership distributions of urban and rural migrants in the supplementary or VIC samples (Table 3) are broadly in line with those in the main samples: both urban and rural migrants are relatively young; homeownership is rare for both groups – in fact in the VIC samples, the homeownership rate for either group is practically nil; and urban migrants are much better educated than rural migrants, although the percentages of post-secondary or higher education attainment for urban migrants in the supplementary samples for both 2005 and 2010 are more than 10 percentage points lower than those reported in the main samples. Living in VICs is apparently not a preferred choice for the better educated urban migrants.

Given the local-migrant divide in terms of housing achievements and socioeconomic backgrounds as well as and the small sample size of rural locals and urban migrants, we combine urban locals and rural locals into the category "locals," and group urban and rural migrants into the category "migrants" for subsequent analysis.

Table 3.   Composition of the supplementary sample.

	2005 Supplementary sample		2010 Supplementary sample	
	Urban mig.	Rural mig.	Urban mig.	Rural mig.
No. of obs.	*71*	*228*	*63*	*241*
Age				
≤20	–	6	6	14
21–30	33	96	34	92
31–40	27	94	15	66
41–50	9	29	6	55
51–60	2	2	2	11
>60	–	1	–	3
Education				
Illiterate	1	5	–	16
Primary school	2	23	4	38
Junior secondary	18	106	12	121
Senior secondary	32	81	29	60
Tertiary degree	13	11	16	2
College	5	–	2	4
Above college	–	–	–	–
Marital status				
Never married	19	53	28	72
Ever married	52	175	35	168
Ownership				
Owner	1	0	3	4
Other	70	228	60	235

Source: Computations based on survey data.

Changing patterns of residential mobility, Guangzhou 1990–2010

Residential mobility rates

We first compute the mobility rate over the period from 1990 to 2010 by pooling the data from both 2005 and 2010 surveys. Here, only residential moves that took place *within* the city of Guangzhou were considered, excluding inter-urban or rural-urban migration which is beyond the scope of this research. To take account of the slightly larger size of the 2010 sample, a weight of 0.49 is applied to the 2005 survey data and 0.51 to the 2010 survey data.[3] We subdivide the entire study period into 10.4 two-year periods, in view of the fact that the 2010 survey was undertaken in the last quarter of that year and hence covered only approximately 0.8 year for the period 2010–2011. The use of two-year rather than one-year periods helps reduce errors arising from random fluctuations. The annual mobility rate is given by the number of moves over a two-year period divided by the number of moving candidates (defined as those over 18 years old who reside in the city in a given period and include both intra-city movers and non-movers), and then further divided by two except for the period 2010–2011, which is divided by 0.8. Figure 1 plots the trend of the mobility rate for migrants and locals in the main sample and also the trend of migrants in VICs.

A few observations are evident from these charts. First, for the local population, the residential mobility rate increased steadily from slightly less than 4 percent per annum in 1990–1991 to over 8 percent per annum in 2000–2001, when the disposal of work-unit housing through heavily discounted sale was at its height. However, full-scale housing commodification with the ending of the welfare allocation of housing under the 1998 housing reform did not bring further increase in the mobility rate. Instead, beginning from the turn of the century the mobility rate exhibited steady declines, and by the end of the 2000s it fell back to the level prevailing in the early 1990s. One explanation for the decline in the mobility rate after 2000 is that after attaining homeownership through purchasing reformed housing in the late 1990s and early 2000s, moving became a more difficult proposition in view of the heavier moving costs under homeownership. Moreover, housing price in Chinese cities, Guangzhou included, increased drastically since the mid-2000s (Hui and Yue 2006) and rendered moving up the housing ladder

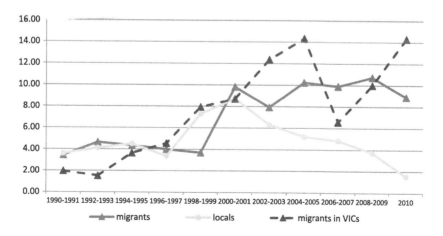

Figure 1.   Residential mobility rates in Guangzhou migrants vs. locals.
Source: Computations based on 2005 and 2010 Guangzhou surveys.

prohibitively expensive; hence, the rather precipitous drop in the mobility rate after 2006–2007.

Second, the trends for migrants depicted by the main sample and by the VIC supplementary sample are quite similar, although in some periods the mobility rate of the former is higher than the latter, and in some other years it was the reverse. Both samples show a steady rising trend from the beginning to the end of the study period, with the mobility rate increasing from below 4 percent in 1990–1991 to over 10 percent in 2008–2009. Neither graph contains a turning point, unlike the trend of locals.

Apparently, migrants in VICs and those elsewhere in the city are confronted with quite similar constraints in making residential decisions. To most non-local households, particularly rural migrants, renting has continued to be the only tenure mode after the full implementation of the 1998 housing reform, under which people with the local *hukou* were busy buying reformed housing. Subsequent conferment of full property rights to owners of reformed housing means that those who have moved up the housing ladder are now able to put up the vacated units in the rental market. Because of this, housing opportunities available to migrants, whether in terms of number or geographical coverage, have been enlarged. This could be a reason behind the continuing rising mobility rates for rural migrants throughout the 2000s. Of course, an equally plausible explanation has to do with the rapid rise in housing rent in recent years, which renders rural migrants with meager means in a permanent state of moving and searching for affordable housing. Moreover, large-scale redevelopment of inner-city neighborhoods and former work-unit compounds as well as of VICs where low-rent housing is concentrated also contributes to the continuing rising mobility rate for this group.

The above conjecture is confirmed by analyzing the change in living space upon residential moves. It can be seen from Table 4 that moving for urban locals was more likely to be associated with an increase in living space in both the 1990s and 2000s than otherwise; however, for rural as well as urban migrants, a move accompanied by reduction in living space was much more likely than otherwise. To migrants, a higher

Table 4. Change in living space upon move by *Hukou* status.

Hukou status	Change in living space			
	Equal	Upward	Downward	Total
1990–2001				
Rural migrant	19 (28.79%)	17 (25.76%)	30 (45.45%)	66 (100.00%)
Urban migrant	10 (43.48%)	2 (8.70%)	11 (47.83%)	23 (100.00%)
Rural local	16 (59.26%)	5 (18.52%)	6 (22.22%)	27 (100.00%)
Urban local	446 (45.37%)	336 (34.18%)	201 (20.45%)	983 (100.00%)
Total (main sample)	491 (44.68%)	360 (32.76%)	248 (22.57%)	1099 (100.00%)
Migrants in VIC	39 (26.35%)	35 (23.65%)	74 (50.00%)	129 (100.00%)
2002–2010				
Rural migrant	82 (43.16%)	30 (15.79%)	78 (41.05%)	190 (100.00%)
Urban migrant	40 (36.04%)	16 (14.41%)	55 (49.55%)	111 (100.00%)
Rural local	6 (27.27%)	5 (22.73%)	11 (50.00%)	22 (100.00%)
Urban local	243 (44.26%)	178 (32.42%)	128 (23.32%)	549 (100.00%)
Total (main sample)	372 (42.55%)	229 (26.26%)	272 (31.19%)	872 (100.00%)
Migrants in VIC	127 (33.96%)	79 (21.12%)	168 (44.92%)	298 (100.00%)

Note: Change of living space per capita is defined as "upward" or "downward" if it increases or decreases by 30percent and is defined as "equal" with a change within 30percent.
Source: Computations based on survey data.

mobility rate does not imply the availability of affordable housing opportunities, whereby they can adjust to changing housing needs by a residential move. Irrespective of the ending of the welfare allocation of housing at the turn of the century, continuing discrimination in the job and housing markets still places severe limits on migrants' housing choice set. In general, they move not in search of a better residence; instead, they move because they are forced to do so.

Determinants of residential moves

In the above we have presented estimates of residential mobility rates over the entire study period of 1990–2010. In this section, we analyze the factors that might affect residential moves and how their effects differ between locals and migrants and vary over time. The retrospective residential and employment histories reported in the surveys enable us to construct a longitudinal data file for every respondent. We divide the whole study period into three main periods; specifically, 1990–1995, 1996–2001, and 2002–2010, with reference to the progress of the housing reform. The first period may be termed early reform; the second reform deepening or housing privatization, with the massive disposal of work-unit housing in association with the 1998 housing reform; and the third post-reform, commodification, when housing was primarily obtained in the market. Discrete-time logit regression is a common modeling approach to analyze time-varying events such as residential moves using longitudinal data files (Allison 1985). We perform six discrete-time logistic regressions – with one regression for households with the local *hukou* in the main sample, and one for migrants by combining the main sample and the supplementary sample – for each of the three time periods. For the migrant models, a dummy variable differentiating the two samples is introduced. Again, we pool the data from both the 2005 and 2010 surveys. For the years from 1990 to 2005, observations from the 2005 survey carry a weight of 0.49 and those from the 2010 survey a weight of 0.51. For the years after 2005, the observations are unweighted, except that observations in the period of 2010 are weighted by a factor of 0.8.

In the regression models, the dependent variable is presence or absence of intra-city residential moves in a given two-year period (yes = 1, no = 0). In the independent variables list, a series of time dummies is included to gage the time trend of the move propensity. According to the graphical analysis given above, for locals the time trend was an increasing one up until the turn of the century; thereafter, the propensity to move declined progressively; for migrants the rate of intra-city mobility has shown a generally rising trend since 1990. The other independent variables belong to two main types. The first refers to the socio-demographic attributes of the household head. Specifically, the variables are as follows:

Age (at the beginning of the two-year period). Young people are generally more mobile, and hence a negative sign is expected. We have tried to include age squared to take account of possible reversion of effect in association with retirement, but finally decided to drop the quadratic term as it proves to be non-significant.

Gender (1 = female; 0 = male). Female-headed households tend to be underprivileged and subject to greater constraints in the choice of housing. But the effect of gender is less clear after controlling for socioeconomic attributes.

Income (personal monthly income). In a redistributive society, which still very much characterized China in early reform times, income *per se* is unlikely to have much influence on housing outcomes. But when housing is primarily accessed in the market, as was the Chinese case in the mid- and late-2000s, income or affordability would be of major importance in defining housing opportunities and hence the move up on the housing ladder. Thus, a positive sign is expected for regressions pertaining to later periods. To facilitate inter-temporal comparisons, relative rather than absolute income is employed. The sample income distribution for each two-year period is divided into 12 levels, and the income level is treated as an interval scale.

Educational attainment. It is measured by eight levels (1 = illiterate; 2 = semiliterate; 3 = primary school; 4 = junior high; 5 = senior high or equivalent; 6 = tertiary education; 7 = college; 8 = above college), and is treated as an interval scale. Education is intimately tied to the possession of redistributive powers in socialist planned economies and hence access to housing opportunities. In market economies, this main ingredient of human capital largely determines a person's position in the job market and hence income. Although through its influence on preference formation, education might continue to exert independent impacts on housing decisions, much of its effects would have been captured by the income variable. The above considerations suggest that the effect of educational attainment on mobility propensity given by the regression models is likely to be higher for the early reform period and smaller in later periods.

Marital status (at the beginning of the two-year period). This is given by: 1 = married and 0 = otherwise. In general, being married tends to inhibit residential mobility, as the move decision involves more than one person. However, marital status correlates with age, and its effect on move propensity may be captured by the latter variable.

Change in marital status (during the two-year period). This is given by 1 = yes and 0 = no. Getting married or experiencing a divorce is an important life event and is likely to trigger a residential move, regardless of the way housing is provided. As such, the variable is expected to have significant positive effects in all time periods analyzed.

Change in job (during the period; 1 = yes; 0 = no). A change in job is a major life-course event that could trigger a move. It is plausible to expect that this variable would assume less importance than change in marital status. However, migrants may respond to the two life-event triggers differently as compared to locals, because the former tend to occupy more temporary jobs and would live close to the workplace to economize on commuting cost. As such, for migrants a change in job could exert stronger effects than a change in marital status. Moreover, given the enlarged opportunity set for individuals to reside near the workplace in a more marketized setting, as it was the case of Guangzhou in the 2002–2010 period, it may be hypothesized that the effect of job change on residential location in this period was larger than in the earlier periods.

Pre-move living space per capita (at the beginning of the two-year period) is a continuous variable to measure crowdedness before a residential move. As housing consumption has emerged as a driving force of intra-city migration in contemporary China, it is expected that crowding would become an important trigger for relocation. Such effect may be more salient in the post-reform era (after 2002) than in the early-reform (before early 1990s) and the reform deepening periods (before early 2000s). However, residential moves in response to crowding may be less visible for migrants than the local population.

Ownership (at the beginning of the two-year period). Again this is a dummy variable, with 1 = owner and 0 = renter. As was pointed out above, in market economies homeownership is negatively associated with residential mobility. But the situation of China in pre-reform and early reform times when work-unit rental housing dominated the housing provision scene would be quite different. Life-long tenure and nominal rent were the norm, and low mobility rates prevailed in the public rental sector.

The second set comprises institutional variables:

Urban hukou. This is a dummy variable used to control for the rural–urban divide in the housing sector, with agricultural or rural *hukou* being the reference category as opposed to non-agricultural or urban *hukou*. Many housing benefits, such as low-rent housing or affordable housing, are tailored to those with the local or urban *hukou* (Wang, Wang, and Wu 2010). In addition, households with the rural *hukou* are more subject to involuntary move due to, for example, insecure employment and redevelopment of informal housing. Hence, those with the rural *hukou* are expected to be more mobile than the urban counterpart.

Economic sector of employment (at the beginning of the period). This variable consists of three dummies pertaining, respectively, to state-owned enterprise, collective enterprise, and the private sector, with government being the reference category. Access to resource in urban China, housing resource in particular, used to depend on the nature of the employment organization. Yet on a priori ground, it is difficult to postulate how this variable would affect residential mobility. The private sector was almost non-existent in the early reform times. Workers of urban collectives were mainly residents of nationalized tenement houses in the inner-city core managed by the municipal housing bureau (Huang 2005); as such, like those in SOEs and government sectors, they also resided in public-sector housing and subject to similar mobility constraints. Accelerated redevelopment of inner-city neighborhoods could result in heightened mobility for workers in urban collectives in more recent times.

Party membership (1 = Chinese Communist Party member; 0 = otherwise). This variable is included to gage the extent to which membership in the CCP affects access to housing opportunities and hence residential mobility, after controlling for socioeconomic and employment status.

We first examine the results for the local population, which are given in Models 1–3 (Table 5). The results generally concur with the hypothesized directions of influence as well as time trends highlighted above. The trends of the move propensity given by the time dummies estimates are broadly in line with the picture given by the bivariate analysis. Major life course events including change in marital status and change in job were significant triggers of residential moves in all three periods. In the case of job change, the effect was larger in 2002–2010 than in earlier periods. Aging is a more gradual event and is correlated with marital status or the family life cycle, arguably the most important factor underlying residential decisions and moves in market economies in the West (Clark 1982). The regression models indicate that the mobility propensity significantly declined with age in the periods 1996–2001 and 2002–2010, with the coefficient estimate of the latter period having a larger magnitude, a result consistent with market deepening. Being married showed significant and positive effects in 1990–1995 when housing was primarily treated as a welfare item, and in 1996–2001 when work units were busy disposing of their housing stock. Educational attainment and income are both indices of socioeconomic status. As hypothesized, the former had a significant positive effect on move propensity in 1990–1995, but its effect diminished in later periods in both magnitude and significance. On the other hand, income was non-significant for 1990–1995, but became highly significant and positive in 1996–2001 and 2002–2010, during which affordability became an increasingly important constraint prohibiting the move up on the housing ladder. Crowding, indicated by pre-move living space per capita, was non-significant in the early-reform era (1990–1995) and became significant and negative, albeit exhibiting only a small effect, in the reform-deepening period

Table 5. Results of discrete time logistic models for locals vs. migrants.

	Models 1–3			Models 4–6		
	Local population in main sample			Migrants in main and VIC samples		
Dependent: intra-city move (0 = no move; 1 = intra-city move)	1990–1995	1996–2001	2002–2010	1990–1995	1996–2001	2002–2010
Age	-0.123	-0.024**	-0.030***	-0.106	0.003	0.006
Female	0.047	0.177	0.405**	1.608	-0.598	0.129
Education	0.224*	0.168*	0.103	0.309	-0.002	0.084
Income	-0.046	0.091**	0.176***	-0.003	-0.001	-0.0004**
Married	0.619*	0.401*	0.167	1.249	0.035	-0.122
Marital status change	1.828***	1.632***	2.040***	-2.705	-0.889	0.773**
Owner	0.112	0.059	-0.795***	-0.495	0.397	-0.348
Job change	0.496[a]	0.699***	0.761**	5.065***	2.818***	1.858***
Living space per capita before move	-0.007	-0.0002[a]	0.012**	0.059**	0.003	0.001
Work sector (reference: government)						
SOE	-0.308	-0.313[a]	-0.337	—	0.229	0.489
Collective enterprise	-0.086	-0.506[a]	0.323	-1.342	-0.545	0.307
Private enterprise	-0.258	-0.152	-0.149	-3.333	-0.603	0.430
Urban hukou (reference: rural)	0.093	0.337	-0.471	—	0.269	-0.068
Party membership	-0.065	0.067	0.354	—	-0.572	1.383***
Migrants in main sample (reference: VIC sample)	—	—	—	-0.447	-0.027	-0.447*
Spell year						
1990–1991	0.235	—	—	-0.094	—	—
1992–1993	0.034	—	—	-0.005	—	—
1994–1995	—	—	—	—	—	—
1996–1997	—	—	—	—	—	—
1998–1999	—	0.861***	—	—	-0.379	—
2000–2001	—	1.146***	—	—	0.923*	—
2002–2003	—	—	—	—	—	—
2004–2005	—	—	-0.213	—	—	0.366*
2006–2007	—	—	-0.744***	—	—	-0.340

	(1)	(2)	(3)	(4)	(5)	(6)
2008–2009	—	—	-0.808^{***}	—	—	0.020
2010	—	—	-3.779^{****}	—	—	-2.379^{***}
Constant	-4.339^{***}	-3.097^{***}	-2.178^{**}	-1.248	-2.642	-2.686^{***}
Model statistics						
Number of obs	2009	3098	5300	211	724	2463
LR X^2	87.37	240.35	323.12	54.38	114.55	238.49
Prob > X^2	0.0000	0.0000	0.0000	0.0000	0.0000	0.0000
Pseudo R^2	0.0731	0.1095	0.1970	0.5353	0.2783	0.1875

Source: Computations based on survey data.
[a] $p < 0.1$; $^*p < 0.05$; $^{**}p < 0.01$; $^{***}p < 0.001$.

(1996–2001). Its effect, however, turned positive and highly significant in the post-reform era. This result should be interpreted in the context of urban China. In the first decade of the twenty-first century, this indicator of crowding probably also measures the size of the windfall gain resulted from discounted home sales. More specifically, people who were able to secure better housing previously enjoyed larger windfall gains from the discounted sale of reformed housing at the end of the 1990s and early 2000s. As such, they were more able to move up the housing ladder in the 2000s. Homeownership only exhibits significant negative effect on moving propensity in 2002–2010,when housing commodification was greatly strengthened.

Regarding the institutional variables, the economic sector of employment only appears to be of importance in the period 1996–2001. That workers in government and related sectors had significantly higher move propensities than those in other economic sectors in this period suggested that they were likely the ones to benefit most in the rush of purchasing discounted reformed housing under the 1998 housing reform. The *hukou* dummy and membership in the CCP are non-significant in all three periods. Their effects may have been captured by other socio-economic factors, such as work sector, education, and income.

Unlike the regressions for the local population, the migrant models (Models 4–6, Table 5) yield few significant variables for all three periods. Income only started to take effect in the more recent period of 2002–2010 – those with higher income were slightly less likely to move probably because they could afford the increased rent demanded by the landlord. Change in marital status exerts smaller influence on a residential move for migrants as compared to the effect of job change, in line with our prediction. For migrants, this variable is only significant for the 2002–2010 period – this it may be because migrants were disposed to moving frequently regardless of change in marital status. In contrast, job change appears to be consistently significant, and the size of its effect is larger than that for the local population. Notably, in all periods residential moves did not respond to pre-move living space per capita. In other words, crowding is not an important motivation for migrants' intra-city moves. Other socio-demographic indicators (e.g. age, gender, and marital status), which showed significant impacts on the local population's moving propensity, were non-significant for migrants. In terms of institutional factors, only Communist Party membership is associated with higher residential motility for migrants in the later periods. No significant differences are found between urban and rural migrants. Additionally, migrants in the VIC sample tend to move more frequently than those in the main sample in the post-reform era (2002–2010), which could be indicative of the heightened redevelopment of VICs in Guangzhou in this period. These results clearly suggest that migrants are subject to forces that are quite different from those of locals in exercising residential decisions and contemplating move. This finding echoes Chan's (1999) observation at the macro-level that employment is the major force behind inter-provincial migrations of non-*hukou* migrants.

Conclusion and discussion

Drawing on residential history data from two surveys conducted in 2005 and 2010, this paper analyzes the trends of residential mobility of Guangzhou residents over the period 1990–2010. In light of the large share of migrants in the population, especially those with the agricultural *hukou* who are subject to discrimination of all kinds, comparison is made between migrants and local residents. The residential mobility rate for Guangzhou local residents increased steadily in the 1990s, but the increasing trend was reversed after the turn of the century with the massive disposal of work-unit and other public-sector housing to sitting tenants upon the end the welfare allocation of housing in 1998. For migrants in both VICs and elsewhere in the city, however, the mobility rate continued to increase in the 2000s. Large-scale redevelopment in the 2000s of VICs as well as inner-city neighborhoods and old work-unit compounds where most cheap rental housing was located likely contributed to the continuing rising mobility trend.

Discrete-time logit analysis employing retrospective longitudinal data of residential and employment histories confirms that major life-course events such as change in marital status and job change are important triggers for residential moves for the local population. Aging is associated with lower mobility propensity, especially in more recent periods. The results also point toward increasing effects of income (positive) and homeownership (negative), and reduced effects of government agency affiliation (change from negative and significant to non-significant) on the mobility propensity over time. In a sense, the results indicate that the causes behind residential moves in Guangzhou increasingly resemble those in market economies in the West.

However, the regressions pertaining to migrants yield quite different results. While for urban locals, a large proportion of moves are for searching for a better residence – either in respect to ownership attainment or increase in living space, for migrants job change appears to be the single-most important trigger of moves. For the latter, most socio-demographic variables, including marital status, show little influence. Income has exerted only marginal effects in recent years. Urban and rural migrants do not show substantial differences in terms of the propensity to move. Despite the neoliberal rhetoric of privatization, commodification, and marketization and the implied more-leveled playing fields, the results confirm once again the disadvantageous positions of the "floating" population in the urban housing market. Institutional barriers confronting migrants are formidable. It is especially true for rural migrants who have been deprived of citizenship rights in their current place of domicile, restricting them to marginal and precarious jobs and largely substandard housing in dilapidated inner-city tenements and VICs. Most migrants remain passive agents in the urban housing market. They may move quite frequently, but this is likely to be a result of unstable employment, eviction by landlord, or redevelopment, as well as sudden adverse change in health and financial conditions (F. Wu 2004). For the great majority of migrant workers in China's leading metropolises, the costs and requirements stipulated by the municipal government for attaining local *hukou* status are beyond reach (Li, Li, and Chen 2010). Thus, for them residential mobility may better be described as being imposed than a choice.

Notes

1. Migrants in this paper refer to non-local migrants moving from one city to another without changing *hukou* status. Chan, Liu, and Yang (1999) have illustrated the importance of distinguishing this group from *hukou* migrants and non-migrants in terms of socioeconomic characteristics and migratory patterns.
2. VIC in China refers specifically to migrant enclaves located in the city center or urban fringe, which provide informal housing for low-income populations, in particular migrants in the city. It is a unique urban phenomenon in China resulting from rapid urbanization, the rural-urban dual land system, and massive rural-to-urban migration. It differs substantially from the New Urbanism-guided village-like communities in the UK or ethnic enclaves in the US Some literature also refers to a VIC as an "urban village," "cheng zhong cun," or "urbanized village". See more discussion in Wang, Wang, and Wu (2010) and Zhu (2014).

3. We use a pooled approach (see O'Muircheartaigh and Pedlow 2002) to combine the two independent samples. The weights are computed in proportion to the relative effective sample size of each survey. The smaller weight for 2005 indicates the smaller sample size in the 2005 survey and a smaller population of Guangzhou in 2005.

References

Alba, Richard D., and Victor Nee. 1997. "Rethinking Assimilation Theory for a New Era of Immigration." *International Migration Review* 31: 826–874.

Allison, Paul D. 1985. *Event History Analysis: Regression for Longitudinal Data*. California, CA: Sage.

Cai, Fang, and Kam Wing Chan. 2009. "The Global Economic Crisis and Unemployment in China." *Eurasian Geography and Economics* 50: 513–531.

Chan, Kam Wing. 1999. "Internal Migration in China: A Dualistic Approach." In *Internal and International Migration: Chinese Perspectives*, edited by F. Pieke and H. Mallee, 49–71. Richmond, VA: Curzon Press.

Chan, Kam Wing, Ta Liu, and Yunyan Yang. 1999. "Hukou and Non-hukou Migrations in China: Comparisons and Contrasts." *International Journal of Population Geography* 5: 425–448.

Chan, Kam Wing, and Li Zhang. 1999. "The Hukou System and Rural-urban Migration in China: Processes and Changes." *China Quarterly* 160: 818–855.

Clark, William A. V. 1982. "Recent Research on Migration and Mobility – A Review and Interpretation." *Progress in Planning* 18: 5–56.

Clark, William A. V., and Frans M. Dieleman. 1996. *Households and Housing Choice and Outcomes in the Housing Market*. New Brunswick, NJ: Center for Urban Policy Analysis, Rutgers University.

Fu, Qiang, and Nan Lin. 2013. "Local State Marketism: An Institutional Analysis of China's Urban Housing and Land Market." *Chinese Sociological Review* 46: 3–24.

Guangzhou Statistical Bureau. 2011. *Guangzhoushi 2010 diliuci quanguo renkou pucha zhuyao shuju gongbao* [2010 Guangzhou Census Bulletin]. Guangzhou: Guangzhou Statistical Information.

Guangzhou Statistical Bureau. 2013. *Guangzhou tongji xinxi shouce 2013* [Manual of Guangzhou Statistics Information 2013]. Guangzhou: Guangzhou Statistical Information.

He, Shenjing, and Fulong Wu. 2007. "Socio-spatial Impacts of Property-led Redevelopment on China's Urban Neighborhoods." *Cities* 24: 194–208.

He, Shenjing, and Fulong Wu. 2009. "China's Emerging Neoliberal Urbanism: Perspectives from Urban Redevelopment." *Antipode* 41: 282–304.

Huang, Youqin. 2005. "From Work-unit Compounds to Gated Communities: Housing Inequality and Residential Segregation in Transitional Beijing." In *Restructuring the Chinese Cities: Changing Society, Economy and Space*, edited by Laurence J. C. Ma and Fulong Wu, 192–221. London: Routledge.

Huang, Youqin, and F. Frederic Deng. 2006. "Residential Mobility in Chinese Cities: A Longitudinal Analysis." *Housing Studies* 21: 625–652.

Hui, Eddie Chi Mun, and Shen Yue. 2006. "Housing Price Bubbles in Hong Kong, Beijing and Shanghai: A Comparative Study." *Journal of Real Estate Finance and Economics* 33: 299–327.

Kain, John F. 1968. "Housing Segregation, Negro Employment, and Metropolitan Decentralization." *Quarterly Journal of Economics* 82: 175–197.

Kain, John F. 1992. "The Spatial Mismatch Hypothesis: Three Decades Later." *Housing Policy Debate* 3: 371–460.

Li, Si-Ming. 2004. "Life Course and Residential Mobility in Beijing, China." *Environment and Planning A* 36: 27–43.

Li, Si-Ming. 2005. "Residential Mobility and Urban Change in China." In *Restructuring the Chinese City: Changing Society, Economy and Space*, edited by Laurence J. C. Ma and Fulong Wu, 157–171. London: Routledge.

Li, Si-Ming. 2010. "Mortgage Loans as a Means of Home Finance in China: A Comparative Study of Guangzhou and Shanghai." *Housing Studies* 25: 857–876.

Li, Si-Ming, and Huiming Du. 2014. "Residential Change and Housing Inequality in Urban China in the Early Twenty-first Century." In *Housing Inequality in Chinese Cities*, edited by Huang Youqin and Li Si-Ming, 18–36. Oxon: Routledge.

Li, Limei, Si-Ming Li, and Yingfang Chen. 2010. "Better City, Better Life, but for Whom? The Hukou and Resident Card System and the Consequential Citizenship Stratification in Shanghai City." *Culture and Society* 1: 145–154.

Li, Si-Ming, and Yat-Ming Siu. 2001. "Residential Mobility and Urban Restructuring under Market Transition-a Study of Guangzhou, China." *The Professional Geographer* 53: 219–229.

Li, Si-Ming, Donggen Wang, and Fion Yuk-ting Law. 2005. "Residential Mobility in a Changing Housing System: Guangzhou, China, 1980–2001." *Urban Geography* 26: 627–639.

Li, Si-Ming, and Zheng Yi. 2007. "The Road to Homeownership under Market Transition Beijing 1980–2001." *Urban Affairs Review* 42: 342–368.

Logan, R. Logan, Yanjie Bian, and Fuqin Bian. 1999. "Housing Inequality in Urban China in the 1990s." *International Journal of Urban and Regional Research* 23: 7–25.

National Bureau of Statistics of China. 2010. *2010 nian quanguo nongmingong jiance diaocha baogao* [Survey Report on Nation-wide Rural-migrant Workers 2010]. Beijing: National Bureau of Statistics of China.

O'Muircheartaigh, Colm, and Steven Pedlow. 2002. "Combining Samples vs. Cumulating Cases: A Comparison of Two Weighting Strategies in NLSY97." *ASA Proceedings of the Joint Statistical Meetings*, New York, NY, 2557–2562.

Owusu, Thomas Y. 1999. "Residential Patterns and Housing Choices of Ghanaian Immigrants in Toronto, Canada." *Housing Studies* 14: 77–97.

Renaud, Jean, and Karine Bégin. 2006. "The Residential Mobility of Newcomers to Canada: The First Months." *Canadian Journal of Urban Research* 15: 67–81.

Rosenbaum, James E., Lisa Reynolds, and Stefanie Deluca. 2002. "How Do Places Matter? The Geography of Opportunity, Self-efficacy and a Look inside the Black Box of Residential Mobility." *Housing Studies* 17: 71–82.

Rossi, Peter H. 1955. *Why Families Move: A Study in the Social Psychology of Urban Residential Mobility*. Glencoe, IL: Free Press.

Solinger, Dorothy J. 1993. "China Transients and the State – A Form of Civil Society." *Politics & Society* 21: 91–122.

Song, Yan, Yves Zenou, and Chengri Ding. 2008. "Let's Not Throw the Baby out with the Bath Water: The Role of Urban Villages in Housing Rural Migrants in China." *Urban Studies* 45: 313–330.

Wang, Ya Ping. 2000. "Housing Reform and Its Impacts on the Urban Poor in China." *Housing Studies* 15: 845–864.

Wang, Ya Ping, Huiming Du, and Li Si-Ming. 2014. "Migration and Dynamics of Informal Housing in China." In *Housing Inequality in Chinese Cities*, edited by Youqin Huang and Li Si-Ming, 87–102. Oxon: Routledge.

Wang, Ya Ping, and Alan Murie. 1996. "The Process of Commercialisation of Urban Housing in China." *Urban Studies* 33: 971–989.

Wang, Ya Ping, and Alan Murie. 2000. "Social and Spatial Implications of Housing Reform in China." *International Journal of Urban and Regional Research* 24: 397–417.

Wang, Ya Ping, and Alan Murie. 2011. "The New Affordable and Social Housing Provision System in China: Implication Comparative Housing Studies." *International Journal of Housing Policy* 11: 237–254.

Wang, Ya Ping, Yanglin Wang, and Jiansheng Wu. 2010. "Housing Migrant Workers in Rapidly Urbanizing Regions: A Study of the Chinese Model in Shenzhen." *Housing Studies* 25: 83–100.

Wang, Feng, and Xuejin Zuo. 1999. "Inside China's Cities: Institutional Barriers and Opportunities for Urban Migrants." *American Economic Review* 89: 276–280.

Wilson, William J. 1987. *The Truly Disadvantaged: The Inner City, the Underclass and Public Policy*. Chicago: University of Chicago Press.

Wu, Fulong. 1996. "Changes in the Structure of Public Housing Provision in Urban China." *Urban Studies* 33: 1601–1627.

Wu, Fulong. 2004. "Intraurban Residential Relocation in Shanghai: Modes and Stratification." *Environment and Planning A* 36: 7–25.

Wu, Weiping. 2004. "Sources of Migrant Housing Disadvantage in Urban China." *Environment and Planning A* 36: 1285–1304.

Wu, Weiping. 2006. "Migrant Intra-urban Residential Mobility in Urban China." *Housing Studies* 21: 745–765. doi:10.1080/02673030600807506.

Yushu, Zhu, Qiang Fu, and Qiang Ren. 2014. "Cross-city Variations in Housing Outcomes in Post-reform China: An Analysis of 2005 Micro Census Data." *Chinese Sociological Review* 46: 26–54.

Zhu, Jieming. 1999. "Local Growth Coalition: The Context and Implications of China's Gradualist Urban Land Reforms." *International Journal of Urban and Regional Research* 23: 534–548.

Zhu, Yushu. 2014. "Spatiality of China's Market-oriented Urbanism: The Unequal right of Rural Migrants to City Space." *Territory, Politics, Governance* 2: 194–217.

Participation and expenditure of migrants in the illegal lottery in China's Pearl River Delta

Zhiming Cheng[a], Russell Smyth[b] and Gong Sun[c]

[a]School of Accounting, Economics and Finance, The University of Wollongong, Wollongong, New South Wales, Australia; [b]Department of Economics, Monash University, Melbourne, Victoria, Australia; [c]Department of Marketing, Central University of Finance and Economics, Beijing, China

Using a unique data-set from the Pearl River Delta in China, we examine the factors associated with rural-urban migrants' participation in and expenditure on illegal gambling. The characteristics that have the largest marginal effects on participation and expenditure are gender, whether one also participates in the legal lottery and playing mahjong and other card games. Of the variables specific to the migration experience, we find that living in a factory dormitory increases the probability of participating in the illegal lottery, and that migrants who are unwilling to give up their farmland are less likely to participate in the illegal lottery. We also find that having a network of female friends is negatively correlated with participation in the illegal lottery. However, other variables specific to migrants, such as willingness to obtain a local *hukou* and remittances, were found to have an insignificant effect on participation.

1. Introduction

The underground economy, which is that part of a country's economic activity that is not officially recorded and is untaxed by the government, is becoming increasingly important in developing and transition economies (Schneider 2005). On the plus side, the underground economy fosters entrepreneurship and provides jobs, goods, and services, especially for the poor (Venkatesh 2006). It can also assist in cutting through bureaucratic red tape and provide an "alternative path" to foster economic development (De Soto 1990). The downside of the underground economy is that it is often associated with illegal activities such as prostitution, drug trafficking, and illegal gambling, which pose significant challenges to economic and social governance (Dreher and Schneider 2010; Schneider and Enste 2000).

Illegal gambling is an important component of the underground economy. Investigating illegal numbers gambling in the US, Sellin (1963, 12) asserts that

... the numbers racket, among syndicated underworld operations, is one of the most highly organized [and] well-staffed, and [is] thoroughly disciplined. Playing the numbers and

63

gambling in general, like participation in other vices, survive because large portions of the population enjoy them, see no harm in them, and do not regard them as immoral, even when they are illegal.

This is still true for the numbers racket in the US today, with an estimated almost US $10 billion annually wagered illegally on the Super Bowl alone (CNBC 2009). Illegal gambling is particularly prevalent in developing and transition countries in Asia. For example, in the late 1990s, illegal gambling constituted as much as 60 percent of the underground economy and was associated closely with organized crime in Thailand (Phongphaichit, Piriyarangsan, and Treerat 1998).

Following China's economic reform, the central government abandoned the Maoist legal norm that prohibited all types of gambling. This resulted in a resurgence in gambling as a preferred form of entertainment among the Chinese (Loo, Raylu, and Oei 2008). In 2012, the nationwide sales of state-run lotteries in China, which were established in 1987, were worth 262 billion renminbi (RMB) (US$42 billion) (Ministry of Finance 2013). Currently, China's legal lottery market is ranked only behind the US lottery market, which was worth US$60 billion in 2012 (Rabinovitch 2013).

Operating alongside the legal lottery market in China is a flourishing illegal lottery market. An illegal lottery called *liuhecai* first appeared in Guangdong province in southern China in the late 1990s. The illegal lottery uses the numbers drawn in an official lottery with the same Chinese name in Hong Kong (called the Mark Six Lottery in English). In recent years, the illegal lottery has spread to and become very popular in all regions of Mainland China as well as Taiwan (Yen and Wu 2013).

China's rural-urban migration has been described as the largest migration flow in history (Zhao 1999). In 2011, more than 160 million rural-urban migrants lived and worked in urban China (Chan 2012); and many of them are active consumers of gambling products (Sun 2012). In China, the *hukou* (household registration) system assigns every Chinese either a rural or urban status. Since rural-urban migrants still hold rural *hukou*, they are not officially recognized as urban citizens. Past studies on rural-urban migrants have studied various aspects of their labor market experience as well as other socioeconomic outcomes (e.g. Cheng 2014; Nielsen, Smyth, and Zhai 2010; Wang and Fan 2012; Zhao 1999). However, there is no research on the extent to which rural-urban migrants participate in the illegal lottery; despite anecdotal evidence suggesting gambling is important in their lives (Gan and You 2013; Sun 2012). For example, a recent survey of 182 migrant workers suggests that approximately one-third had participated in lotteries and 45 percent had engaged in other gambling games after work; moreover, some rural-urban migrants took leave to gamble immediately after receiving their monthly salary (Gan and You 2013).

An individual's willingness to take risks may be related to the migration decision. A risky decision involves choice among a range of possible outcomes whose probabilities are known (Tversky and Kahneman 1992). This is epitomized in betting in which the probabilities are known. Thus, there is an analogy between the migration decision and gambling (Williams and Balaz 2012). Hence, migrants risk preferences with respect to gambling might be informed by their risk preferences with respect to migration. The relationship, however, between risk preferences and the decision to migrate is ambiguous (Jaeger et al. 2010; Williams and Balaz 2012). On the one hand, migration can be understood as risky behavior, because of uncertainty about future wages, living conditions, relationships with family and friends in the place of origin, and cultural adjustment (Williams and Balaz 2012, 2014). On the other hand, migration might be seen as risk ameliorating on the basis that nonmigration also contains risk/uncertainty and that migration helps to spread risk across uncorrelated markets (Stark and Bloom 1985; Stark 1991). The limited empirical evidence on the relationship between risk aversion/tolerance and migration suggests that individuals who are more willing to take risks are more likely to migrate (Barsky et al. 1997; Dohmen et al. 2005; Heitmueller 2005; Jaeger et al. 2010). There are also demographic differences between migrant groups. For example, Balaz and Williams (2011) found that female migrants were much more risk tolerant than female nonmigrants, while risk tolerance between male migrants and nonmigrants was similar. Research suggests that in China, migrants also exhibit high-risk tolerance with respect to having unprotected sex in light of the risk of HIV infection (Li, Morrow, and Kermode 2007).

The purpose of this study is to examine the determinants of participation in and expenditure on the illegal lottery among rural-urban migrants. To do so, we employ a unique representative data-set collected in Guangdong province – one of the major destinations for rural-urban migrants, the birthplace of the illegal lottery, and the largest provincial market for both the legal and illegal lotteries.

We extend the literature on migration and illegal gambling in the following important ways. First, despite the growing prevalence of illegal gambling in China, there is little research on it and no statistical studies on the determinants of participation in illegal gambling. Second, there are no statistical studies on the extent to which migrants participate in illegal gambling or gambling of any kind. Third, most existing studies focus exclusively on the determinants of participation in and expenditure on lotteries. In addition to examining determinants of participation and expenditure, we also examine the frequency with which rural-urban migrants participate in the illegal lottery.

Focusing on the problem of illegal gambling, especially in the rural-urban population, is important for the following reasons. First is the size of the problem. It is estimated that one trillion RMB (US $161 billion) is wagered illegally every year in China (Eimer 2010). There are no official statistics on the numbers that play the illegal lottery, but Bosco, Liu, and West (2009) speak in terms of an illegal "lottery fever" that has spread through southern China, which literally brings large numbers to a standstill on the nights in which the lottery is drawn in Hong Kong. Second, because it is illegal, the illegal lottery implicitly represents a challenge to the authority of the Chinese communist government. This applies a fortiori to participation of rural-urban migrants in the illegal lottery, given that the Chinese government is particularly concerned about minimizing social unrest in this segment of the population.

Third, rural-urban migrants generally face poor living and working conditions and experience low quality of life (e.g. Gao and Smyth 2011; Knight and Gunatilaka 2010; Nielsen, Smyth, and Zhai 2010). Many studies have found that rural-urban migrants feel like "outsiders" in China's urban areas (e.g. Zhuang 2013). Male migrants, in particular, often turn to drinking and gambling on the illegal lottery to escape their sense of loneliness and in hope of improving their lot in life. There is evidence that illegal gambling among rural-urban migrants is associated with myriad negative outcomes, such as betting addiction and lottery-related crime. The illegal lottery in China has been linked to organized crime, including criminal illegal activities such as drugs and prostitution (Human Rights in China 2002; Liang 2009). One study estimates that there are seven million "problem gamblers" participating in illegal lotteries in China, of which, 430,000 are seriously addicted (Evans 2012). There are several reports of cases where rural-urban migrants have committed crime in order to finance addiction to playing the illegal lottery (e.g. Cui 2013). A better understanding of the factors determining the frequency with which rural-urban migrants participate in the illegal lottery and the amount spend on the illegal lottery will assist policy-makers in formulating intervention strategies that minimize the impact of problem gambling behavior and its implications among migrants and other populations.

2. The illegal lottery in China

Punters in the Chinese illegal lottery bet on the special number drawn in the Hong Kong Mark Six Lottery. There are 49 balls, so the winning probability is 1/49, or 2 percent, with a payout rate typically at 40-to-1. In addition to betting on the special number, a series of sideline bets are also available. For example, punters can bet on the color of the special number (red, green, or blue), the Chinese zodiac signs pre-assigned to each of the lottery balls, whether or not the special number is odd or even, whether or not the special number is larger than 25, the last digit of the special number, and so on. There is no minimum or maximum amount of a given bet. The odds vary according to the specific side bet.

A peculiar aspect of the illegal lottery in China is that numerous versions of "hint sheets" with revealing photos of women, cartoon figures, classic poetry, riddles, and jokes are published and widely circulated by illegal publishers (Figure 1 for an example).

Figure 1. A sample hint sheet.
Notes: A sample hint sheet for No. 28 Draw on March 12, 2013. Punters believe that information on the special number is contained somewhere on the sheet. The special number for this draw was 4; however hints related to almost all 49 numbers can be found on the sheet. Punters believe that if they do not win it is because they are not smart enough to find the correct information or else it is divine will. The note at the top right corner shows that this sheet is published by the spurious "Hong Kong Rich Women Network" and the note at the bottom shows the contacts for Mr Zeng, a non-existent Taoist prophet at "Hong Kong Mark Six Information Center" who helps people get rich.

Punters also seek enlightenment through the popular media. For example, many believe that the British BBC pre-school children's television series Teletubbies contains hints revealed by prophets, making it popular among punters. The belief that hint sheets contain winning numbers has led many punters to imagine that lotteries, which are random in nature, represent, in essence, a predictable game and profitable investment (Liu 2011).

The illegal lottery operates within a pyramid structure with four tiers. The bottom tier consists of the punters. The next tier up consists of those who collect and record bets from the punters. Many bet collectors, some of whom only accept bets from punters with whom they are familiar, are either the proprietors or employees at corner stores, newsagents, eateries, or internet cafes. This facilitates easy and covert contact with punters. Bet collectors usually receive 10 percent of the bet as commission. The bet collectors pass the bets to small bookies, which represent the second pyramid tier from the top. At the top of the pyramid are the larger bookies who take some of the larger bets from the smaller bookies. Following the draw, the bookies give the winnings to the bet collectors to distribute to the punters. A credit account is available for trusted repeat punters and big punters, in which they can settle their accounts following the draw. In general, there is a high level of trust between the stakeholders in the illegal lottery (Deng 2006). Many Chinese prefer the illegal lottery to the official lottery, because there have been a number of instances of cheating in the legal lottery, and punters believe the lottery in Hong Kong to be fairer.

A feature of the illegal lottery pyramid structure is that participants only have contact with those at the tier immediately above them. Thus, the punters do not have contact with the small bookies, and the bet collectors have no contact with the large bookies. In most cases, communication is top down and one way between big bookies, small bookies, and bet collectors via mobile phones and instant messaging programs. The Chinese government has failed to eradicate the illegal lottery despite periodic crackdowns on the punters and bet collectors, because the secretive tiered structure provides protection to both the small and the large bookies, who are financing the illegal lottery further up the pyramid. This feature of the illegal lottery also helps explain why the illegal lottery flourishes. High-stake gamblers and larger bookies are protected by the police and party officials, at least in popular perception.

In Hong Kong, the Mark Six Lottery is broadcast on television on Tuesday, Thursday, and Saturday. Since the broadcast is blocked in Mainland China, internet and modern communication technology play important roles in marketing the illegal lottery and publicizing the winning numbers. Some enterprises have also attempted to benefit from the illegal business. Before 2005, one of China's biggest domestic search engines provided a large number of paid advertisements for illegal lottery websites. Later, these advertisements were banned at the request of the internet administrators. However, currently one of the largest international search engines still provides abundant information linking to illegal lottery websites. Because of the large volume of cell phone traffic immediately before and after the draw (around 8 to 9 pm on draw nights), it is often difficult for people to connect on their cell phones during this period. In 2006, one of the authors observed in a city in Guangdong province that a state-owned mobile service provider erected outdoor advertising that proclaimed: "Choose us! Your cell phone call will be able to get through at the critical moments on Tuesday, Thursday, and Saturday nights!"

At the beginning of the century, the illegal lottery was predominately confined to the rural areas of China, but it has now spread and is widely followed in both rural and urban China (Zeng 2004 for a detailed account of its early development). There is no national statistic on the magnitude of the illegal lottery in China, but some statistics demonstrate its impact in particular locales. Reflecting its roots, the illegal lottery is most popular in southern China. In the rural areas of Guangdong province, it is estimated that 3.3 billion RMB was spent on the illegal lottery in 2003 and that 40 percent of the expenditure flowed to bookies (Liang 2004). In Hunan, Yueyang City, which has a population of 5.3 million, more than 300 million RMB flowed to bookies in 2004. In Miluo City (also in Hunan), which has a population of 650,000, more than 5 million RMB was remitted to bookies on every lottery draw day, and the amount of personal bank savings decreased by 97 million RMB in December 2004 alone (Guo 2005). In northern China, in one village of 6000 households near Tianjin, it was reported that 6 million RMB was spent on the illegal lottery within a half year, and local bank savings decreased by 500 thousand RMB each month in 2003 (Wen 2004). One of the peaks of betting occurred in December 2004, when a red special number had not been drawn for 19 consecutive draws in the Mark Six Lottery. In view of the high probability that a red special number would be drawn, nearly 20 billion RMB was spent by punters during that month alone (Li 2005).

3. Literature review

3.1. International research on lotteries

Theoretical and empirical research on the economics of lotteries has been plentiful (Ariyabuddhiphongs 2011; Clotfelter and Cook 1990; Perez and Humphreys 2013). The majority of these studies have focused on mature legal lottery markets (such as Australia, Spain, the United Kingdom, and the US). Most of the extant research has examined the relationships between demographic and socioeconomic characteristics and gambling behavior. Several variables have been tested for a correlation with participation in and expenditure on mostly legal lotteries. Some general themes emerge in the literature. In this section, we review the variables considered in previous studies in order to provide a general overview before adding more contextualized variables for our analysis of China's illegal lottery.

Generally speaking, males are more likely to participate in lotteries than females (Coups, Haddock, and Webley 1998; Herring and Bledsoe 1994). But, some studies have reached the opposite conclusion. In Sydney, Australia, females had a higher preference for lotteries than males, and they experienced problem gambling at levels comparable to males (Hing and Breen 2001). Most studies have found that those who are single are more likely to participate in lotteries and be more frequent gamblers than those who are married (Hodgins et al. 2012). In Malaysia, household size has been found to have a statistically negative effect on lottery participation and expenditure among Chinese households (Tan et al. 2010).

Differences in age and life-cycle patterns may lead to variations in lottery participation and expenditure. But the relationship between age and gambling is uncertain. Barnes et al. (2011) find that the frequency of participation in lotteries in the US exhibits a nonlinear relationship in which participation is at a maximum between one's 30s and 60s. But in some states, such as Massachusetts, the 65-and-over age group are the major participants in lotteries and are the main age group responsible for increasing per capita sales of the state-run lottery (Jackson 1994). Gambling among youth is also of particular concern. Felsher, Derevensky, and Gupta (2003) find that youth participate in all forms of lotteries, because their parents purchased tickets for them, and that young participants with significant gambling problems had much higher parental participation in the lottery compared with nongamblers and social gamblers.

Most studies find that the better educated are less likely to play lotteries (Coups, Haddock, and Webley 1998; Mikesell 1991; Stranahan and Borg 1998b). Similarly, a number of studies find that income is negatively correlated with participation in lotteries. Lotteries are particularly popular among low-income individuals in the US who may substitute lottery play for other entertainment or view lotteries as a means to improve their standard of living (Clotfelter and Cook 1991). Blalock, Just, and Simon (2007) find a strong and positive relationship between participation in lotteries and poverty rates, but no relationship with movie ticket sales, another inexpensive form of entertainment, and poverty rates. However, findings for income are not uniform. Scott and Garen (1994) find that, controlling other demographic and socioeconomic variables, income has no significant effect on expenditure on lotteries. Stranahan and Borg (1998a) suggest that income has a negative and significant impact on participation in lotteries, but no significant effect on lottery expenditure, conditional on participation. As a consequence, Humphreys, Lee, and Soebring (2011) suggest that estimates of the income elasticity of demand for lotteries are susceptible to model selection and specification and call for more empirical attention to this issue.

Other common findings are that participation in lotteries is positively correlated with peer participation in lotteries, frequency of participation in other forms of gambling, and misunderstanding about the probability of winning (Coups, Haddock, and Webley 1998). Frequency of participation in lotteries has been found to be positively correlated with scratch card play, gambling on horse and greyhound racing, participating in the football pools, and bingo as well as with beliefs about skill, luck, and optimism (Rogers and Webley 2001). Findings on the relationship between religion and gambling in Western contexts have been mixed. Some studies have found that the religious are more likely to gamble and that their intensity of gambling is higher (Hodge, Andereck, and Montoya 2007; Hoffmann 2000; Joukhador, Blaszczynski, and Maccallum 2004), while other studies find an insignificant, or even negative, relationship between being religious and gambling (Diaz 2000; Lam 2006). Among higher frequency gamblers, individuals with obsessive-compulsive disorder, higher anxiety or depression, and with higher impulsivity and antisocial personality traits have been found to be more likely to have self-reported gambling-related problems (Hodgins et al. 2012).

3.2. *Research on the illegal lottery in China*

The literature on gambling in Mainland China is scant, and most is focused on legal lotteries. A limited number of studies provide descriptive details on the characteristics of those who participate in legal lotteries. One such study, using data from Guangzhou in Guangdong, shows that the majority of participants in the official lottery were young, middle-aged males with above-average income and a high-school education or better (Zeng and Zhang 2007). Another such study finds that most participants in legal online lotteries in China are well educated and married with relatively high average income (approximately 3000–5000 RMB per month) (Chen 2012). Still other studies have examined problem gambling behavior associated with the official sports lottery (Li et al. 2012a) and the relationship between the official lottery administration and lottery consumption (Li et al. 2012b).

There are two studies in Chinese that examine determinants of participation in lotteries in mainland China using a regression framework. One study by Zhang and

Zheng (2006) examines lottery participation among 550 people in Haikou city in Hainan province. Employing a probit model, they find that being male, married, less educated, and having lower income are correlated with participation in lotteries. Another study, using data collected from Yangzhou city in Jiangsu province, finds that neither marital status nor income were significantly correlated with lottery participation (Zhang 2007). A limitation of both studies is that neither distinguishes between participants in legal and illegal lotteries and thus do not provide insights into the characteristics of those who participate in the illegal lottery.[1]

In the English literature, there are a few anthropological studies focusing on gambling in the rural areas. Bosco, Liu, and West (2009) point to an increasing level of social acceptance of illegal lotteries in rural China. They argue that the illegal lottery mirrors and mimics the ability of the consumer capitalist economy to produce wealth through control of key nodes in the economy rather than through production. The illegal lottery, they argue, captures the alchemy of neoliberalism, which is "to yield wealth without production, value without effort" (Comaroff and Comaroff 2000, 313), and it is an example of "millennial capitalism – that odd fusion of the modern and the post-modern, of hope and hopelessness, of utility and futility, of promise and its perversions" (283).

Based on long-term fieldwork in Hubei province, Steinmüller (2011) argues that popular gambling reflects how boundaries of what is socially acceptable are changing in the Chinese countryside.[2] Steinmüller's work, however, largely de-genders gamblers and depicts gamblers as mostly male petty capitalists. To provide a more nuanced gendered picture of gambling, Gong (2013) analyzes how women justify their participation in the illegal lottery in rural China and suggests that women's participation in the illegal lottery is tied to seeking peer approval. Due to data limitations, however, Gong (2013) was unable to provide a statistical profile of the percentage of females participating in the illegal lottery. According to a survey in 2007 by Zeng and Zhang (2007), nearly 80 percent of punters in legal lotteries in Guangzhou were males. But little is known about the gender profile in the illegal lottery.

Some of these studies mention the role of religion and superstition as factors underpinning participation in illegal lotteries and the strategies punters use to select colors and numbers (see e.g. Bosco, Liu, and West 2009; Liu 2011). However, in these studies, the relationship between religion and superstitious factors, on one hand, and participation in the illegal lottery is only briefly touched upon and not quantified.

To summarize, the existing literature has focused on describing the characteristics of those who participate in either the official lotteries in Mainland China or fail to distinguish between the legal and illegal lotteries. Those studies that do explicitly focus on the illegal lottery take an anthropological perspective and are centered on understanding aspects of the illegal lotteries in rural China. There are no studies that examine determinants of participation in illegal lotteries in urban China and no studies that consider the involvement of rural-urban migrants in either the legal or illegal lotteries.

4. Data and descriptive statistics

The data used in this study came from the Survey of Rural-to-Urban Migrant Workers in the Pearl River Delta, which was administered in nine prefecture-level cities (*dijishi*) by Sun Yat-Sen University in 2006. The survey was a general purpose survey, designed to elicit response about socioeconomic and living conditions of migrant workers. A team from Sun Yat-Sen University approached the employers of migrant workers. To ensure that the data were representative of the rural-urban migrant population, the sampling framework adhered to three principles. First, the allocation of the sample across the nine cities followed the actual proportional distribution of the migrant population in these cities according to the 2010 official population census. Second, the allocation of the sample across secondary and tertiary industries followed the actual proportional distribution of migrants in these industries according to the official statistical yearbook in 2005. Third, no more than three interviewees were recruited from each employer.

The survey contained several questions that elicited information of interest to this study. These questions included whether the respondent participated in the illegal lottery (i.e. *liuhecai*) and, if so, how often he or she participated,[3] the amount of the average bet the respondent placed, the total number of bets if the respondent had ever won, the total number of times he or she had won, and the highest winning prize. In addition, respondents were asked whether they participated in official lotteries and additional questions in relation to their after-work lifestyle, such as if they played mahjong and card games, which usually involve gambling in the Chinese context. Other information collected in the survey pertained to income, employment status, and personal and family characteristics. Respondents were also asked, on a three-point or four-point scale about a series of subjective indicators, such as their concept of identity, the extent to which they had adapted to urban life, the extent to which life was difficult, and the extent to which they felt a sense of social injustice. Only those respondents who answered the lottery-related questions are included in the analysis, making the final sample size 3937.[4] The distribution of the sample across cities is presented in Table 1.

Table 1. Sample distribution.

Tier of city*	City	Frequency	Percent
First tier	Guangzhou	1285	32.64
	Shenzhen	755	19.18
Second tier	Zhuhai	194	4.93
	Foshan	269	6.83
	Dongguan	609	15.47
	Zhongshan	197	4.98
Third tier	Zhaoqing	197	5.00
	Huizhou	200	5.08
	Jiangmen	232	5.89
	Total	3937	100.00

*Tier of city is defined by the Bureau of Human Resources and Social Security of Guangdong Province according to the minimum wage level and socioeconomic development.

Table 2. Participation in the illegal lottery.

	Frequency	Percent
No	3207	81.40
Yes	730	18.60
Formerly	311	42.60
Occasionally	326	44.66
Frequently	67	9.18
Constantly	26	3.56
Total	3937	100.00

Table 3. Descriptive statistics.

Variable		Number of observations	Mean (std. dev.)/ percentage	Min	Max
Illegal lottery					
Won prize		730	0.7027 (0.4574)	0	1
No	(=0)	217	29.72		
Yes	(=1)	513	70.28		
Normal amount of a bet (RMB)		717	46.29 (142.59)	1	2000
Total amount of bets (RMB)		643	3399.25 (13273.55)	1	180,000
Amount of the largest prize (RMB)		502	1372.89 (4703.85)	1	60,000
Total amount of prizes (RMB)[a]		585	1219.51 (4097.91)	0	50,000
Demographic, socioeconomic and employment characteristics					
Gender		3937			
Female	(=0)	1756	44.60		
Male	(=1)	2181	55.40		
Marital status		3935			
Unmarried	(=0)	1914	48.64		
Married	(=1)	2021	51.36		
Age		3934	29.59 (9.94)	14	71
<20	(=0)	541	13.75		
20–29	(=1)	1582	40.21		
29–39	(=2)	1040	26.44		
39–49	(=3)	548	13.93		
>49	(=4)	223	5.67		
Schooling (years)		3934	9.24 (2.27)	6	15
Primary and below	(=0)	871	22.14		
Junior high	(=1)	1980	50.33		
Senior high	(=2)	592	15.05		
Technological school	(=3)	385	9.79		
Graduate diploma	(=4)	106	2.69		
Monthly income (RMB)		3760	1101.20 (678.20)	0	20,000
<700	(=0)	618	16.44		
700–1000	(=1)	1210	32.18		
1000–1300	(=2)	1006	26.76		
1300–1600	(=3)	466	12.39		
>1600	(=4)	460	12.23		
Religious		3937			
No	(=0)	2872	72.29		
Yes	(=1)	1101	27.71		
Household size		3934	4.94 (1.77)	1	20
Have child(ren)		3937			
No	(=0)	1937	50.11		
Yes	(=1)	1964	49.89		
Employment type		3937			
Formal	(=0)	3062	77.77		
Informal	(=1)	875	22.23		
Migrate with children?		1964			
No	(=0)	1425	72.56		
Yes	(=1)	539	27.44		
Migrate with spouse?		1420			
No	(=0)	137	9.65		

(*Continued*)

Table 3.   (*Continued*).

Variable		Number of observations	Mean (std. dev.)/ percentage	Min	Max
Yes	(=1)	1266	89.15		
Number of job changes		3899	1.89 (2.39)	0	30
Number of male friends in city		3756	4.33 (9.42)	0	250
Number of female friends in city		3736	2.85 (7.76)	0	280
Live in factory dormitory?		3936			
No	(=0)	2256	57.32		
Yes	(=1)	1680	42.68		
Vote in village committee election?		3928			
No	(=0)	3565	90.76		
Yes	(=1)	363	9.24		
Willing to give up farm land?		3917			
Yes	(=0)	1024	26.14		
No	(=1)	1599	40.82		
Unsure	(=2)	1294	33.04		
Willing to obtain local *hukou*?		3929			
Yes	(=0)	1635	41.61		
No	(=1)	1365	34.74		
Unsure	(=2)	929	23.64		
Monthly expenses (RMB)		3840	586.06 (420.46)	0	5000
Monthly remittances (RMB)		3688	257.65 (373.90)	0	5833.333
After-work activities					
Read newspaper and book		3937			
No	(=0)	2213	56.21		
Yes	(=1)	1724	43.79		
Study					
No	(=0)	3911	99.34		
Yes	(=1)	26	0.66		
Do housework		3973			
No	(=0)	3109	78.97		
Yes	(=1)	828	21.03		
Look after child(ren)					
No	(=0)	3925	99.70		
Yes	(=1)	12	0.30		
Play mahjong and cards		3973			
No	(=0)	3319	84.3		
Yes	(=1)	618	15.7		
Participate in official lotteries		3973			
No	(=0)	3665	93.09		
Yes	(=1)	272	6.91		
Subjective indicators					
Social identity[b]		3937			
Peasant	(=0)	2451	62.26		
Nonpeasant	(=1)	737	18.72		
Ambiguous/overlapping	(=2)	749	19.02		
Labor rights violated		3059			
No	(=0)	2331	76.20		
Yes	(=1)	728	23.80		
Adaptation to urban life[c]		3826			
Poor	(=0)	240	6.27		
Neutral	(=1)	1644	42.97		

(*Continued*)

Table 3. (*Continued*).

Variable		Number of observations	Mean (std. dev.)/ percentage	Min	Max
Good	(=2)	1942	50.76		
Life is difficult		3976			
Never	(=0)	1453	38.28		
Occasionally	(=1)	1488	39.20		
Often	(=2)	741	19.52		
Always	(=3)	114	3.00		
Feel sense of social injustice		3699			
Never	(=0)	1285	34.74		
Occasionally	(=1)	1290	34.87		
Often	(=2)	706	19.09		
Always	(=3)	418	11.30		
Feel sense of income injustice		3627			
Never	(=0)	1520	41.91		
Occasionally	(=1)	1252	34.55		
Often	(=2)	613	16.90		
Always	(=3)	241	6.64		
Lack workplace belonging		3723			
Never	(=0)	1721	46.23		
Occasionally	(=1)	1226	32.93		
Often	(=2)	480	12.89		
Always	(=3)	296	7.95		

Notes: [a]Includes some participants who had never won a prize; [b]"Social identity" refers to migrant workers' answer to the question: "What is your social identity in your own opinion?' [c]"Adaption to urban life" refers to migrant workers' answer to the question: "How well do you adapt to urban life?".

Tables 2 and 3 present descriptive statistics for the sample. Just over half the sample is male (55.4 percent) and married (51.4 percent). The average age is 29.6 years. The education level is low, with almost three quarters (72.5 percent) educated to only junior high school or below. The participation rate in the illegal lottery was 18.6 percent. This was significantly higher than the participation rate in official lotteries (6.9 percent). Among 730 participants in the illegal lottery, approximately 70 percent had won at least once in the lottery. The normal amount of a bet ranged from 1 to 2000 RMB, reflecting the flexible bet requirement. The average normal amount of a bet was 46 RMB, while the total amount wagered by any one individual ranged from 1 to 180,000 RMB. The average "normal bet" consumed about 4 percent of mean monthly income, which was around 1100 RMB. If the individual played all 12–13 games in a month, nearly half of their mean monthly income would be spent on the illegal lottery.

5. Econometric methods

In the 1980s and early 1990s, some studies employed standard tobit models unconditional on lottery participation to examine lottery expenditure. This approach assumes that independent variables have the same impact on probability of participation and expenditure. However, it may lead to biased estimates, because lottery consumption is subject to two decisions: whether to bet, and, if so, how much to bet. For example, studies in the US find that ethnicity/race does not affect participation, but African-Americans, Hispanics, and other minorities tend to spend more if they do gamble (Scott and Garen 1994; Stranahan and Borg 1998a). Another study finds that while African-American youth are less likely than average to gamble, they tend to do so more frequently if they do gamble (Welte et al. 2001).

There are two major approaches with similar methodological frameworks to address the presence of nonparticipants in the data-set. One is the double-hurdle model, which has been used in modeling usage of and expenditure on tobacco and alcohol. The other is a two-stage model, which distinguishes between the decision to participate in the lottery and the decision of how much to spend on the lottery. The two-stage approach entails using a probit model to estimate lottery participation and then a truncated tobit model to estimate lottery expenditure (Stranahan and Borg 1998b). We adopt the two-stage approach, because it permits a more flexible specification in selecting variables than the double-hurdle model. In the first stage of the estimation, this entails modeling the probability of participation in the illegal lottery through a binary outcome model. Equation (1) is of the form:

$$y_i = \beta x_i + \varepsilon_i \tag{1}$$

where y_i^* is the ith respondent's expenditure on the illegal lottery, x_i is a vector of explanatory variables affecting participation in the illegal lottery, β is a vector of parameters, and ε_i is the error term. Equation (1) can be converted to a probit model of illegal lottery participation of the form:

$$\text{Prob}(y_i > 0) = \Phi[\beta x_i] \text{Prob}(y_i > 0) = 1 - \Phi[\beta x_i] \tag{2}$$

where Φ is the standard normal cumulative distribution function. Equation (2) implies that the respondent participates in the illegal lottery when actual expenditure is positive.

The second stage estimates the actual expenditure conditional on participation, meaning that expenditure is truncated at zero. A truncated tobit model models a sample of respondents who had a positive level of expenditure on the illegal lottery. The truncated tobit model is of the form:

$$\emptyset(y_i|y_i > 0) = \frac{1}{\Phi}\left(\frac{\beta x_i}{\sigma}\right)\left[\left(\frac{1}{\sqrt{2\pi\sigma}}\right)\exp\left\{-\left(\frac{1}{2\sigma^2}\right)(y_i - \beta x_i)^2\right\}\right] \tag{3}$$

where $\emptyset(y_i|y_i > 0)$ is the standard normal density of y_i conditional on participation in the illegal lottery and $\emptyset(y_i) \sim N(\beta x_i, \sigma^2)$.

While most existing studies only consider whether or not respondents had participated in gambling, the survey also provides information on migrants' frequency of participation in the illegal lottery. Table 2 suggests that 45, 9, and 4 percent of migrant workers participated on an occasional, frequent, and constant basis, respectively. This allows us to consider the probability of a respondent falling into each frequency category. A multinominal logit model considers the effect of the vector of explanatory variables x_i on the probability that individual i will fall into in one of the j categories of participation (v_{ij}):

$$\text{Prob}(v_{ij}) = \frac{\exp(\beta_j x_i)}{\Sigma_{k=1}^{j} \exp(\beta_k x_i)} \quad \text{for } j = 1, \dots, 5 \tag{4}$$

With the multinomial model nonparticipants act as the reference group, and all coefficients are normalized to zero. Maximum likelihood is employed in estimating Equations (2)–(4).

6. Results

Table 4 shows the marginal effects from a probit regression for determinants of participation in the illegal lottery. Model 1 includes basic demographic and socioeconomic variables. Males are 14.6 percent more likely than females to participate in the illegal lottery. While this finding is consistent with most previous research, the gap between the probability of males and females participating in the illegal lottery is significantly much smaller than in legal lotteries, where nearly 80 percent of punters are males (see e.g. Zeng and Zhang 2007). This is consistent with our field findings that the illegal lottery is more popular among females than the legal lottery. A possible explanation is that participating in legal lotteries, which often requires a visit to lottery selling booths or stores, is often regarded by females as a "man's game," because males are the overwhelmingly majority of customers. This is also consistent with females being more likely to feel shame and guilt for gambling, such as buying legal lotteries publicly (McCormack, Shorter, and Griffiths 2014). The convenience of betting on the illegal lottery, via undercover bet collectors on a one-to-one basis, makes females more willing to participate. In addition, this convenience is of particular importance for migrant workers who have a busy work schedule.

The marginal effects for age and age squared show an inverted U-shaped pattern, implying that migrant workers are initially more likely to participate in the illegal lottery as age increases, but the probability of participation decreases after age 29. Marital status, having children and schooling all have an insignificant effect on participation. Migrants from larger households are more likely to participate in the illegal lottery. For each additional member of the household, participation increases 0.8 percent. This result may imply that migrant workers from larger households face more financial and/or life pressure in the cities and that they are more attracted by the illegal lottery with a stronger desire to win. Migrant workers with higher income are more likely to participate, but the marginal effect is small. To examine this point, we grouped income into different categories in Models 2–5, discussed below.

Having religious beliefs significantly increases the probability of participating in the illegal lottery. Those who state that they are religious are 5.7 percent more likely to participate than those who are not religious. This result is consistent with the notion that superstitious/religious ideas associated with the illegal lottery and supposedly embedded in hint sheets and in cryptic messages in the popular media have more credence with people who have religious beliefs. As Binde (2007, 145) puts it:

> Gambling and religion have certain elements in common: notions of the unknown, mystery, and fate, as well as imagery of suddenly receiving something of great value that changes life for the better. In many traditional cultures gambling has existed in concord with polytheistic and animistic religion; gambling and religion go well together precisely because of the elements they have in common.

In China and many traditional non-Western societies, in particular, religious and superstitious beliefs are often interwoven with participation in gambling. Chinese people believe that one's lot in life can be changed if it is the will of the gods and that it is possible to turn the tide of fortune through seeking divine intervention to help them win the lottery (Tse, Rossen, and Wang 2010). Consequently, Chinese gamblers often pray to the gods for success and explain wins and losses in terms of divine will (Binde 2007; Harrell 1974).

Table 4. Determinants of participation in the illegal lottery, based on a binary probit regression.

Variable	(1) Marginal effect	(2) Marginal effect	(3) Marginal effect	(4) Marginal effect	(5) Marginal effect	(6) Marginal effect	(7) Marginal effect
Male	0.1461 (11.12) ***	0.1377 (10.37) ***	0.1423 (6.38) ***	0.1186 (6.03) ***	0.1247 (8.28) ***	0.0892 (6.45) ***	0.1368 (8.49) ***
Age	0.0245 (4.68) ***		0.0008 (-0.07)	0.0219 (2.96) **	0.0221 (3.87) ***		0.0193 (3.00) **
Age²/100	-0.0366 (-5.12) ***		-0.00074 (-0.57)	-0.0313 (-2.92) **	-0.0315 (-4.26) ***		-0.0294 (-3.17) **
Married	-0.0179 (-0.63)	0.0080 (0.45)		-0.0374 (-1.05)	-0.0262 (-0.92)	0.0120 (0.68)	0.0069 (0.31)
Schooling	0.0011 (0.38)	0.0001 (0.04)	0.0067 (1.37)	0.0023 (0.59)	0.0020 (0.27)	0.0008 (0.28)	0.0023 (0.63)
Income	0.00003 (3.85) ***						0.00004 (3.33) ***
Religious	0.0569 (3.97) ***	0.0587 (4.10) ***	0.0895 (3.53) ***	0.0459 (2.47) *	0.0441 (3.00) *	0.0574 (4.01) ***	0.0537 (3.13) ***
Household size	0.0079 (2.31) *		0.0097 (1.63)	0.0102 (2.32) *	0.0083 (2.34)		
Have child(ren)	0.0240 (0.80)		0.0407 (0.96)	0.0627 (1.57)	0.0390 (1.27)		
Informal job	-0.0377 (-2.12) *						
Tier of city (ref: first tier)							
Second tier	-0.0301 (-2.05) *		-0.0257 (-1.04)	-0.0155 (-0.85)	-0.0137 (-0.95)		
Third tier	-0.0320 (-1.74)		-0.0636 (-2.16) *	-0.0166 (-0.74)	-0.0203 (-1.11)		
Income group (ref: <700)							
700-1000		0.0222 (1.05)				0.0218 (1.03)	
1000-1300		0.0648 (2.90) *				0.0518 (2.33) *	
1300-1600		0.1073 (3.91) ***				0.0863 (3.18) **	
>1600		0.1078 (3.81) ***				0.0736 (2.69) **	
Age group (ref: <20)							
20-29		0.0585 (2.55) *				0.0504 (2.19) *	
29-39		0.0988 (3.30) **				0.0700 (2.37) *	
39-49		0.0239 (0.75)				0.0114 (0.36)	
>49		-0.0141 (-0.38)				0.0016 (0.04)	
Migrate with children			0.0252 (1.00)				
Migrate with spouse			-0.0103 (-0.29)				
Number of job changes				0.0071 (2.31) *	0.0082 (3.34) **		
Number of male friends in city				0.0018 (1.62)	0.0030 (-0.20)		
Number of female friends in city				-0.0076 (-2.98) **			
Five or more male friends in city					-0.0464 (-2.37) *		
Live in factory dormitory				0.0269 (2.83) *	0.0243 (2.70) *		
Vote in village committee election				0.0059 (0.27)	0.0137 (0.64)		
Willing to give up farm land (ref: yes)							
No				-0.0557 (-3.46) **	-0.0546 (-3.49) ***		
Unsure				-0.0392 (-2.35) *	-0.0332 (-2.05) *		
Willing to obtain local hukou (ref: yes)							
No				-0.0193 (-1.25)	-0.0146 (-0.98)		
Unsure				0.0166 (0.95)	0.0157 (0.93)		
Expenses				0.00008 (5.05) ***	0.00007 (4.61) ***		
Remittances				-2.16e-07 (-0.15)	-5.18e-09 (-0.00)		
After-work activities							
Reading						-0.0355 (-2.37) **	
Study						-0.0323 (-0.41)	
Do housework						-0.0157 (-1.00)	
Look after children						0.0103 (0.09)	
Play mahjong and card games						0.1618 (8.90) ***	
Participate in official lottery						0.3408 (11.77) ***	

(Continued)

Table 4. (Continued).

Variable	(1) Marginal effect	(2) Marginal effect	(3) Marginal effect	(4) Marginal effect	(5) Marginal effect	(6) Marginal effect	(7) Marginal effect	
Subjective indicators								
Social identity (ref: peasant)								
Nonpeasant							0.0018 (0.11)	
Ambiguous/overlapping							0.0388 (2.23)	*
Labor rights violated							0.0327 (1.81)	*
Adaptation to urban life (ref: poor)								
Neutral							0.0689 (1.83)	
Good							0.0547 (1.47)	
Life is difficult (ref: never)								
Occasionally							0.0265 (1.51)	
Often							0.0758 (3.05)	*
Always							0.0608 (1.04)	
Social injustice (ref: never)								
Occasionally							0.0394 (2.06)	*
Often							0.0427 (1.70)	
Always							0.0982 (2.99)	**
Income injustice (ref: never)								
Occasionally							-0.0173 (-0.98)	
Often							0.0109 (0.47)	
Always							-0.0143 (-0.43)	
Lack workplace belonging (ref: never)								
Occasionally							-0.0363 (-2.11)	*
Often							-0.0243 (-0.97)	
Always							-0.0293 (-0.92)	
Log likelihood	-1675.9722	-1677.6312	-653.71856	-1508.8579	-1578.1225	-1547.1147	-1088.0299	
LR χ^2	244.26	240.95	82.09	251.80	262.80	501.98	197.79	
Prob>χ^2	0.0000	0.0000	0.0000	0.0000	0.0000	0.0000	0.0000	
Number of observations	3752	3752	1399	3354	3354	3752	2512	

Note: z-statistics in parentheses.
*$p < 0.05$; **$p < 0.01$; ***$p < 0.001$.

Migrant workers in an informal job are 3.4 percent less likely to participate in the illegal lottery than those in a formal job. One possible reason for this result is that the collective living arrangements, such as factory dormitories and shared rental housing, provide formal workers with a more active and familiar gambling environment, as well as easier access to bet collectors near residential areas. Another possible reason is that formal workers have a higher income and better benefits and with more stable employment are less concerned about spending on the illegal lottery than informal workers.

Using more nuanced specifications, Models 2–5 provide more information on how grouped income, grouped age, and after-work activities effect participation. In Model 1, income has a significant, but a very small marginal effect. Each additional 100 RMB of income increases the likelihood of participation by 0.33 percent.[5] Therefore, in Model 2, income is categorized into five groups. Compared with the lowest income group (earning less than 700 RMB per month), respondents earning 1000–1300 RMB per month are 6.5 percent more likely to participate, respondents earning 1300–1600 RMB per month are 10.7 percent more likely to participate, and those earning more than 1600 RMB per month are 10.8 percent more likely to participate. This result suggests that the illegal lottery "tax" exhibits a progressive trend and imposes a greater burden relative to income on the relatively rich than on the relatively poor among migrant workers. This differs from the evidence of a regressive legal lottery tax found in many other countries (Hansen 1995; Price and Novak 2000), but is similar to the case of Spain, where, in contrast to previous research on the legal lottery in developed countries, a strong relationship was observed between legal lottery expenditure and income, with an estimated income elasticity higher than one (Perez and Humphreys 2011). While legal lotteries can be viewed as implicit taxes contributing to government revenue from a public finance perspective (Clotfelter and Cook 1987), in the illegal lottery in China, it is the bookies that have benefited from participation of higher income punters.

Model 2 also examines the relationship between age and participation in the illegal lottery through exploring potential nonlinear effects. The results suggest that migrant workers in their twenties and thirties are the most likely to participate in the illegal lottery. Relative to migrants aged less than 20, migrants in their twenties are 5.9 percent more likely to participate in the illegal lottery and those aged in their thirties are 9.9 percent more likely to participate in the illegal lottery.

Models 3 and 4 explore the relevance of various migration variables. Migrating with children and migrating with spouse has no effect on participation in the illegal lottery. The marginal effect of expenses on the probability of participating in the illegal lottery is small, while remittances have a statistically insignificant effect on participating in the illegal lottery. Number of job changes has a weakly statistically significant effect (at the 10 percent level) on participating in the illegal lottery, but the marginal effect is small. To be specific, an individual in the sample with the average number of job changes (1.89) has a 1.3 percent higher likelihood of engaging in the illegal lottery.

In terms of network effects, the number of male friends is insignificant. Each additional female friend reduces the likelihood of engaging in the illegal lottery by 0.76 percent. A migrant with the average number of female friends (2.85) is 2.2 percent less likely to participate in the illegal lottery. We also used a specification in Model 5 in which instead of the number of male and female friends, we used binary variables for five or more male and female friends. The results suggest that having five or more male friends is not significant, but having five or more female friends reduces the likelihood of participation by 4.7 percent, which is significant at the 5 percent level. Thus, having a large network of male friends has no effect on participation, but having a large group of female friends (more than five) has a negative effect on participating in the illegal lottery. One explanation for this finding is that it is more likely that women will have a large network of female friends than men, and women are less likely to participate in the illegal lottery. A second explanation is that having a large network of female friends may present migrants with more leisure alternatives to gambling. A third possibility is that having a large network of female friends ameliorates some of the causes of gambling, such as isolation and loneliness. Fourth, women are more risk averse than men (see e.g. Jianakoplos and Bernasek 1998). Thus, there may be peer pressure from having a large network of female friends to gamble less.

Individuals who are not willing to give up farmland are 5.6 percent less likely to participate in the illegal lottery, relative to those who are willing. Willingness to give up farmland can be seen as a proxy for risk aversion with individuals who are not willing to give up farmland being more risk averse. Stark (1991) argues that migration represents a mechanism through which the household can spread risk, in which the migration decision is a self-enforcing cooperative contractual arrangement. In Stark's framework, household members spread risk because they are assumed to be risk averse. Retaining farmland is consistent with taking a broader household perspective in which some members of the household continue to farm the land. Participation in lotteries can take up some, or even a large component, of a migrant's income, meaning that there is less income available for remittance to household members in the countryside, resulting in breach of the contractual arrangement. That risk-averse migrants are less willing to give up farmland and have a lower probability of gambling, renders supports for Stark's (1991) theory of a family strategy of risk spreading through migration.

Migrants living in factory dormitories are 2.7 percent more likely to participate in the illegal lottery. There are several possible explanations for this result. First, for migrants living in factory dormitories, there may be a limited range of alternative leisure activities. Second, living conditions in factory dormitories are often poor (Smith and Pun 2006). In particular, long working hours and lack of work-life divide associated with living on site in a factory dormitory can result in psychological problems, which are associated with a higher proclivity to engage in gambling (Smyth et al. 2013). In several studies, gamblers have reported that gambling represents an avenue, albeit temporarily, to distract them from problems and worries (Petry 2005). This idea is a central element of psychological models of gambling (see e.g. Sharpe 2002). Third, the factory dormitory has been used as a means by which employers can control migrant workers. At the same time, factory dormitories in China have become sites for migrant resistance and rallying points for lobbying for better working conditions (Smith and Pun 2006). Engaging in an activity which is formally illegal may be seen as an implied form of resistance to authority. Fourth, this can also be related to migrants' network effect on illegal gambling as discussed earlier as well as risk-pooling behavior among roommates within the same social networks (Attanasio et al. 2012). In China, group buying is becoming more and more popular among participants in both legal and illegal lotteries.

In Model 6, in addition to basic demographic variables, a series of variables related to life style are included. Migrant workers who enjoy reading after work are 3.6 percent less likely to participate in the illegal lottery. This result might reflect that avid readers are better informed of the low odds of winning or that reading and playing the lottery are substitutes as leisure activities. Existing research has found that those who are constrained in terms of the time that they have to allocate to leisure are more likely to participate in lotteries as a form of leisure (Casey 2006). Those who enjoy reading after work might have more time to participate in leisure. Other activities, such as studying, doing housework, and looking after children are not correlated with participation in the illegal lottery.

If migrant workers enjoy playing mahjong and card games (usually involving gambling) and participating in the official lottery, they are more likely to participate in the illegal lottery. Those who play mahjong and card games are 16.2 percent more likely to participate in the illegal lottery, while those who participate in the official lottery are 34.1 percent more likely to also participate in the illegal lottery. These results suggest that the illegal lottery is complementary to official lotteries and other forms of gambling for many punters, and that betting on the illegal lottery has become an integral part of gambling culture that views minor gambling as a harmless enjoyable activity (*xiaodu yiqing*). In other words, participation in the illegal lottery may be perceived by migrant workers as entertainment rather than gambling (Yoong, Koon, and Min 2013). This is consistent with US experience, where people who participate in one lottery tend to participate in others as well, regardless of legal status (Hybels 1979).

It is expected that Guangdong will soon experiment with a new official lottery similar to the Mark Six Lottery in a bid to capture some of the market from the illegal lottery. The results here challenge the view that developing a legal version of the Mark Six Lottery will suppress the illegal lottery through diverting punters to the legal lottery market. It is unlikely that a legal Mark Six Lottery in Guangdong would replace the illegal lottery given complementary effects of legal and illegal lotteries. The results also help explain why the illegal lottery has not been eliminated, but has become more popular despite numerous police raids during a period of rapid expansion of the legal lottery market in China.

In urban China, migrants struggle in reshaping their self-identity, while facing institutional barriers such as the *hukou* system (Wang and Fan 2012). Many of them, especially younger migrants, have not engaged in agricultural work, have little emotional attachment with the countryside, and have a desire to become recognized as urban citizens. Nonetheless, being defined as peasants in the *hukou* system and as urban workers in the process of urban industrialization, rural-urban migrants have ambiguous and overlapping identities (Pun 1999). This has resulted in a state of limbo for migrant workers in which "there is no future as a laborer [while] returning to the village has no meaning" (Chan 2010, 659). Moreover, many urban locals "look down" on rural-urban migrants, making it challenging for the latter to adapt to urban life (Nielsen and Smyth 2008). Migrant workers often have strong feelings of social and economic injustice (Davis and Feng 2008) and impingement of their legal rights is prevalent (Chan 2001). Table 3 presents the descriptive statistics of such subjective indicators of wellbeing.

In Model 7, in addition to demographic variables, we include these indicators of subjective well-being. Perceptions of adaptation to urban life and income injustice have no statistically significant effects. Those who self-report having an ambiguous or overlapping social identity are 3.9 percent more likely to participate in the illegal lottery than those who perceive themselves as peasants. Those who consider that their labor rights are violated are 3.3 percent more likely to participate in the illegal lottery. Those who occasionally consider life difficult (as opposed to never) are 2.7 percent more likely to participate in the illegal lottery. Those who occasionally and those who always feel a sense of social injustice are 3.9 and 9.8 percent, respectively, more likely to participate in the illegal lottery relative to those who never feel a sense of social injustice. A possible explanation is that migrant workers who self-report having an ambiguous or overlapping social identity believe their labor rights violated, consider their lives difficult, or feel a sense of social injustice regard winning in the illegal lottery as a chance to make positive change. This is consistent with media observation that the main demand for lotteries comes from the poor and disadvantaged members of society who see winning the lottery as a means to change their lives in the context of deepening income inequality in urban China (Rabinovitch 2013).

One cannot draw inferences about causality between generic subjective indicators, such as feeling a sense of social injustice or feeling that life is difficult, and playing the lottery from the results in Table 4. In the gambling psychology literature, more generally, the direction of causality between subjective well-being and gambling has not been fully established (see the studies reviewed in Tang and Oei 2011). Those who feel a sense of social injustice or that life is difficult might participate in the lottery as a form of escapism or in the hope of winning and improving their lot. Alternatively, playing the lottery (and losing) might contribute to feeling a sense of injustice or feeling that life is difficult.

To explore this issue further, we present bivariate probit instrumental variable (IV) estimates in which we instrument for "life is difficult" and "feel a sense of social injustice." To instrument for "life is difficult," we use positive responses to questions asking respondents whether they were penalized for not holding a valid temporary residence permit (*zanzhu zheng*), whether they need to pay a sponsorship fee (*zanzhu fei*) for their children to attend the local public school, and whether they feel that local people do not trust them. To instrument for "feel a sense of social injustice," we use positive responses to a question asking respondents whether they are willing to participate in collective action to fight for their rights and benefits if violated. Results of a Wald test of exogeneity and other statistical tests (results not shown) suggest that both the bivariate probit IV models are appropriate. The coefficients on the IVs are significant at the 1 percent level with a positive sign in first-stage regressions, suggesting that they are correlated with

Table 5. Bivariate probit IV estimates for "life is difficult" and "feel a sense of social injustice".

Variable	Marginal effect	
Life difficult (no = 0, yes = 1)[a]	0.0515 (1.09)	***
Other control variables?	Included	
Wald test of exogeneity: $\chi^2(1)$	3.19	
Prob > χ^2	0.0105	
ρ	0.0181	**
Number of observations	2767	
Social injustice (no = 0, yes = 1)[b]	0.0821 (1.23)	***
Other control variables?	Included	
Wald test of exogeneity: $\chi^2(1)$	5.49	
Prob > χ^2	0.0192	
ρ	0.0161	**
Number of observations	2642	

Notes: IVs: [a]I was penalized for not holding a valid temporary residence permit (*zanzhu zheng*); I need to pay a sponsorship fee (*zanzhu fei*) for my child to attend the local public school; Local people do not trust me (Yes = 1, No = 0 in each case). [b]I am willing to participate in collective action to fight for my rights and benefits if they are violated (Yes = 1, No = 0). *$p < 0.05$; **$p < 0.01$; ***$p < 0.001$; z-statistics in parentheses.

"life is difficult" and "feel a sense of social injustice," respectively. However, there is no reason to believe that any of the IVs would be correlated with participating in the illegal lottery. The marginal effects for "life is difficult" and "feel a sense of injustice" from the IV estimates are reported in Table 5. Both variables are significant at the 1 percent level, and the marginal effects are similar to Model 6 in Table 4. The marginal effects of the control variables, which are not reported, are also similar to those reported in Model 6 in Table 4.

Table 6 presents results from a truncated tobit regression for the determinants of expenditure on a normal bet and total bets. The same variables are statistically significant regardless of how the dependent variable is measured. The results for age, gender, religion, reading after work, playing card games and mahjong after work, and participation in the official lottery are significant with the expected signs and are similar to the probit for participation in the illegal lottery. Gender, playing cards, and participating in the official lottery have the largest marginal effects on both the size of a normal bet and total bets. The amount of a normal-sized bet placed by males is 30 percent higher than that placed by females, based on the mean value of a normal bet. The corresponding figures for those who participate in the official lottery and those who play mahjong or card games, respectively, are 48 and 39 percent higher. The coefficients on education and marital status

Table 6.   Determinants of expenditure on the illegal lottery, based on truncated tobit regression.

Variable	Normal amount of a bet			Total amount of bets		
	Coefficient		Marginal effect	Coefficient		Marginal effect
Male	67.78 (7.18)	***	14.27	5865.98 (6.64)	***	1189.61
Age	12.24 (3.60)	***	2.58	869.57 (2.75)	**	176.35
Age²/100	−17.98 (−3.82)	***	−3.78	−1291.81 (−2.98)	**	−261.98
Married	−12.59 (−1.01)		−2.65	−37.48 (−0.03)		−7.60
Schooling	0.22 (0.11)		0.05	17.91 (0.10)		3.63
Income	0.02 (3.47)	**	0.004	2.33 (4.73)	***	0.47
Religious	31.64 (3.49)	***	6.66	3369.47 (3.98)	***	678.32
Household size	4.00 (1.77)		0.84	326.20 (1.52)		65.1595
Informal job	−10.54 (−0.95)		−2.22	−1065.79 (−1.03)		−216.14
Read newspaper and book	−19.35 (−2.23)	***	−4.07	−1604.09 (−1.98)	*	−325.31
Play mahjong and card games	85.29 (8.48)	***	17.96	6314.02 (6.64)	***	1280.47
Participate in official lottery)	105.06 (8.03)	***	22.12	11410.61 (9.25)	***	2314.04
Constant	−467.12 (−7.90)	***		−40340.75 (−7.30)	***	
Log likelihood	−5264.3687			−7472.8678		
LR χ²	388.45 (12)			369.79 (12)		
Prob>χ²	0.0000			0.0000		
Number of observations	3740			3667		
Censored observations	3057			3057		
Uncensored observations	683			610		

Note: *t*-statistics in parentheses.
*$p < 0.05$; **$p < 0.01$; ***$p < 0.001$.

Table 7. Determinants of patterns of participation in the illegal lottery, based on multinominal logistic regression.

Variable	Formerly Relative risk ratio		Occasionally Relative risk ratio		Frequently Relative risk ratio		Constantly Relative risk ratio	
Male	1.9313 (4.64)	***	2.0193 (4.71)	***	8.4424 (4.02)	***	3.1265 (1.76)	
Age	1.1510 (2.58)	*	1.1495 (2.45)	*	1.0169 (0.15)		1.2392 (1.13)	
Age^2/100	0.8109 (−2.71)	**	0.8095 (−2.64)	**	0.9647 (−0.25)		0.7559 (−1.06)	
Married	0.7968 (−1.20)		1.2053 (0.96)		1.9307 (1.56)		0.8405 (−0.29)	
Schooling	1.0036 (0.12)		1.0135 (0.44)		1.0670 (1.03)		0.8731 (−1.32)	
Income	1.0002 (1.96)	*	1.0002 (1.98)		1.0000 (−0.19)		1.0003 (2.00)	*
Religious	1.6594 (3.81)	***	1.3874 (2.36)	**	1.1757 (0.53)		1.0459 (0.09)	
Household size	1.0347 (0.99)		1.0498 (1.43)		1.0415 (0.57)		1.1240 (1.17)	
Informal job	0.9113 (−0.55)		0.8786 (−0.77)		1.0977 (0.29)		0.8214 (−0.37)	
Read newspaper and book	0.8990 (−0.78)		0.7257 (−2.37)	**	0.5036 (−2.39)	**	0.2844 (−2.61)	**
Play mahjong and card games	2.3926 (5.91)	***	2.8584 (7.44)	***	2.6602 (3.51)	***	3.9139 (3.17)	**
Participate in official lottery	2.0249 (3.06)	**	7.6407 (11.62)	***	14.6775 (8.97)	***	24.6662 (6.84)	***
Log likelihood	−2235.9082							
LR χ^2	601.15 (48)							
Prob>χ^2	0.0000							
Pseudo R^2	0.1185							
Number of observations	3752							

Notes: z-statistics in parentheses. The reference group is nonparticipants, and all estimated coefficients for this group are normalized to zero.
*$p < 0.05$; **$p < 0.01$; ***$p < 0.001$.

are insignificant, similar to the probit for participation, and while income is significant, the marginal effect is small, similar to Models 1, 3, and 6 in Table 4. However, differing from the results for participation, household size and formality of employment are not statistically significant determinants of expenditure on the illegal lottery.

Table 7 shows the relative risk ratios of falling into different frequency categories from the multinominal logistic regression. There are several interesting points. First, reading after work is not significant in predicting who falls into the former punter category, but reading is inversely related to being an occasional, frequent, or constant punter. Second, higher income increases the probability of being a former or constant punter, but has no effect on being an occasional or frequent punter. Third, being religious increases the probability of being a former or occasional punter, but has no statistical effect on being a frequent, or constant, punter. This result implies that for frequent and constant punters, superstition is less influential in playing the illegal lottery. This is consistent with casual observation that frequent participants in the lotteries prefer to calculate the probability of winning numbers rather than looking for clues hidden in the hint sheets, and that playing the lottery is a normal consumption for some punters. Fourth, playing social gambling games and buying official lottery tickets are the most significant factors increasing the relative risk ratios of falling into each of the frequency categories from former to constant. Finally, when it comes to explaining the characteristics of those who constantly participate in the illegal lottery, the most important predictors are playing card games, participating in the legal lottery, and having a relatively high income.

7. Conclusion

This study has examined migrant participation in and expenditure on the illegal lottery using representative data from the Pearl River Delta in Guangdong. Considering separate decisions in participation and expenditure, we applied a two-stage approach in which a probit model was used to investigate the determinants of participation, and a truncated tobit model was used to examine the determinants of expenditure. We also employed multinominal logit regression to examine the determinants of falling into different gambling frequency categories compared with nonparticipants.

We find that migrant workers who are male, in their twenties or thirties, have relatively high income, are religious, come from larger households, work in the formal sector, participate in the legal lottery, and play mahjong and cards are more likely to participate in the illegal lottery. Of these factors, gender, participating in the legal lottery, and playing mahjong and cards have the largest marginal effects. We also find that living in a factory dormitory increases the probability of participating in the illegal lottery, and that migrants who are unwilling to give up their farmland are less likely to participate in the illegal lottery. Having a (larger) network of female friends in city is negatively correlated with participation in the illegal lottery.

We find that marital status, schooling, and having children had no significant effect. Our results differ from previous findings for China reported in Zhang and Zheng (2006), who find that urban residents who are married, less educated, and have lower income were more likely to participate in lotteries. In contrast to Zhang and Zheng, who do not distinguish between the legal and illegal lotteries, we focus on the illegal lottery, and our sample is restricted to rural-urban migrants.

With a few exceptions, we found that the same factors that were associated with participation in the illegal lottery were also associated with expenditure, conditional on participation. The exceptions were household size and whether the individual was in formal employment, which were insignificant in explaining expenditure. Moreover, the same factors that were associated with the amount of a normal bet were associated with the amount of total bets. The factors that had the largest marginal effects in explaining participation – gender, participation in the legal lottery, and playing mahjong and cards – also had the largest marginal effects in explaining expenditure on the illegal lottery.

In developed countries, psychological intervention and guidance on personal finance for the treatment of gambling problems are widely used. However, implementing these methods would be difficult in China, because gambling (problem and pathological) is usually regarded as a private or household matter rather than a social problem. Some government agencies and NGOs have been working to help rural-urban migrants in various ways, but they are less concerned with the illegal gambling problem. There is much scope to give more attention to assisting rural-urban migrants and other populations affected by the illegal lottery. Our results suggest that intervention policies should target high-risk groups such as males, those in their twenties and thirties, and those with relatively higher income.

There are various ways through which research on the illegal lottery in China can be improved. Data collection should cover different socioeconomic groups to provide a better base for a comparative study, such as between rural residents, rural-urban migrants, and urban locals. Previous research has found that in Taiwan, lottery advertising is more influential on individuals with lower incomes and schooling (Lee and Chang 2008). Due to data limitations, we are unable to examine whether anti-illegal lottery advertising and crackdowns have unintentionally helped the illegal lottery expand through raising people's awareness of the product (Borg and Stranahan 2005). Because of data limitations, we were also unable to examine the influence of other factors, such as personality and impulsive sensation seeking – which have been found to be important in other countries (McDaniel and Zuckerman 2003) – on lottery participation and expenditure among migrant workers. Future research on these facets, with appropriate data, has the potential to generate further results with important policy implications that could help improve the lives of vulnerable groups, such as migrants. More generally, future research could examine the relationship between gambling and some of the major socioeconomic issues facing China. For example, one might study how participation in gambling contributes to poverty and income inequality among migrant workers or in the population as a whole. Mixed methods or qualitative research could also be used to examine the relationship between gambling and poverty in specific cases. This, in turn, has the potential to shed further insights into how activities such as gambling affect the lives of vulnerable, marginalized, groups such as migrants.

Acknowledgements

The authors thank the anonymous referees, Prof. Kam-Wing Chan, Dr Liang Choon Wang, Associate Professor Melanie Beresford, Prof. Yuanliang Song and Weini Liao for their constructive comments and suggestions. Zhiming Cheng thanks the research grant from Macquarie University.

Notes

1. These two papers do not report the sampling method or describe the data collection.
2. Steinmüller's (2011) study mentions the illegal lottery but its main focus is on other forms of gambling especially mahjong.

3. We do not have data on whether the respondents participated in gambling prior to migration and, if so, the extent to which they bet. Thus, we are unable to say anything about whether migrants gamble more or less compared to prior to migrating.
4. The original sample size was 3973. We dropped 33 responses from respondents that did not answer the questions related to the illegal lottery, and three further responses from respondents with dubious answers after cross-checking. The distribution of the dropped responses across different groups of income was very similar to the final sample.
5. The relatively small income effect could possibly translate into a larger effect because migrant workers' labor incomes have increased substantially since 2008.

References

Ariyabuddhiphongs, V. 2011. "Lottery Gambling: A Review." *Journal of Gambling Studies* 27: 15–33.

Attanasio, Orazio, Abigail Barr, Juan Camilo Cardenas, Garance Genicot, and Costas Meghir. 2012. "Risk Pooling, Risk Preferences, and Social Network." *American Economic Journal: Applied Economics* 4: 134–167.

Balaz, Vladimir, and Allan M. Williams. 2011. "Risk Attitudes and Migration Experience." *Journal of Risk Research* 14: 583–596.

Barnes, Grace M., John W. Welte, Marie-Cecile O. Tidwell, and Joseph H. Hoffman. 2011. "Gambling on the Lottery: Sociodemographic Correlates Across the Lifespan." *Journal of Gambling Studies* 27: 575–586.

Barsky, B. Barsky, F. Thomas Juster, Miles S. Kimball, and Matthew D. Shapiro. 1997. "Preference Parameters and Behavioral Heterogeneity: An Experimental Approach in the Health and Retirement Study." *Quarterly Journal of Economics* 112: 537–579.

Binde, Per. 2007. "Gambling and Religion: Histories of Concord and Conflict." *Journal of Gambling Issues* 20: 145–165.

Blalock, Garrick, David R. Just, and Daniel H. Simon. 2007. "Hitting the Jackpot or Hitting the Skids: Entertainment, Poverty, and the Demand for State Lotteries." *American Journal of Economics and Sociology* 66: 545–570.

Borg, Mary O., and Harriet A. Stranahan. 2005. "Does Lottery Advertising Exploit Disadvantaged and Vulnerable Markets?" *Business Ethics Quarterly* 15: 23–35.

Bosco, Joseph, Lucia Huwy-Min Liu, and Matthew West. 2009. "Underground Lotteries in China: The Occult Economy and Capitalist Culture." In *Economic Development, Integration, and Morality in Asia and the Americas*, edited by D. C. Wood, 31–62. Bingley: Emerald.

Casey, Emma. 2006. "Domesticating Gambling: Gender, Caring and the UK National Lottery." *Leisure Studies* 25: 3–16.

Chan, Anita. 2001. *China's Workers Under Assault: The Exploitation of Labor in a Globalizing Economy.* New York: M.E. Sharpe.

Chan, Kam Wing. 2010. "The Global Financial Crisis and Migrant Workers in China: There Is No Future as a Laborer; Returning to the Village Has No Meaning." *International Journal of Urban and Regional Research* 34: 659–677.

Chan, Kam Wing. 2012. "Migration and Development in China: Trends, Geography and Current Issues." *Migration and Development* 1: 187–205.

Chen, Haiping. 2012. "Problem Gamblers and Irrational Purchase of Lotteries in China." Paper presented at the Conference on Problem Gamblers and Responsible Lotteries, Beijing, March 25.

Cheng, Zhiming. 2014. "The New Generation of Migrant Workers in Urban China." In *Urban China in the New Era: Market Reforms, Current State, and the Road Forward*, edited by Zhiming Cheng, Mark Wang, and Junhua Chen, 125–153. Berlin: Springer.

Clotfelter, Charles T., and Philip J. Cook. 1987. "Implicit Taxation in Lottery Finance." *National Tax Journal* 40: 533–546.

Clotfelter, Charles T., and Philip J. Cook. 1990. "On the Economics of State Lotteries." *The Journal of Economic Perspectives* 4: 105–119.

Clotfelter, Charles T., and Philip J. Cook. 1991. *Selling Hope: State Lotteries in America.* Cambridge, MA: Harvard University Press.

CNBC. 2009. "The Big Payoff." December 15. Accessed July 15, 2013. http://video.cnbc.com/gallery/?video=1359692716

Comaroff, Jean, and John L. Comaroff. 2000. "Millennial Capitalism: First Thoughts on a Second Coming." *Public Culture* 12: 291–343.

Coups, Elliot, Geoffrey Haddock, and Paul Webley. 1998. "Correlates and Predictors of Lottery Play in the United Kingdom." *Journal of Gambling Studies* 14: 285–303.

Cui, Ping. 2013. *Trucker Robbed Women in Order to Buy Illegal Lottery.* Accessed April 30. http://www.chinadaily.com.cn/hqgj/jryw/2013-04-09/content_8713693.html

Davis, Deborah S., and Wang Feng, eds. 2008. *Creating Wealth and Poverty in Postsocialist China.* Stanford: Stanford University Press.

De Soto, Hernando. 1990. *The Other Path: The Invisible Revolution in the Third World.* San Francisco, CA: Perennial Library.

Deng, Yanhua. 2006. "Dixia Liuhecai Zai Nongcun Shehui De Yunzuo Luoji: Guanxi Yu Xinren De Yunyong - Yangcun Tianye Yanjiu." [The Operational Principles of Underground Liuhecai in the Rural Context: The Exertion of Guanxi and Trust.] *Shehui* [Society] 1: 167–186.

Diaz, Joseph D. 2000. "Religion and Gambling in Sin-City: A Statistical Analysis of the Relationship between Religion and Gambling Patterns in Las Vegas Residents." *The Social Science Journal* 37: 453–458.

Dohmen, Thomas, Armin Falk, David Huffman, Uwe Sunde, Jurgen Schupp, and Gert G. Wagner. 2005. *Individual Risk Attitudes: New Evidence from a Large, Representative, Experimentally Validated Survey.* IZA Discussion Paper No. 1730. Bonn: Institute for the Study of Labor.

Dreher, Axel, and Friedrich Schneider. 2010. "Corruption and the Shadow Economy: An Empirical Analysis." *Public Choice* 144: 215–238.

Eimer, David. 2010. "China's Secret Gambling Problem." Accessed April 4, 2013. http://www.telegraph.co.uk/news/worldnews/asia/china/6942975/Chinas-secret-gambling-problem.html

Evans, Michael. 2012. *China Catches Lottery Fever.* Accessed April 30, 2013. http://www.aljazeera.com/indepth/features/2012/12/20121223144826895159.html

Felsher, Jennifer R., Jeffrey L. Derevensky, and Rina Gupta. 2003. "Parental Influences and Social Modelling of Youth Lottery Participation." *Journal of Community and Applied Social Psychology* 13: 361–377.

Gan, Mantang, and Yuyang You. 2013. "Bad Habits of the New Generation Migrant Workers and Intervention Strategy of Industrial Social Work." *Journal of South China Agricultural University* 12: 105–111.

Gao, Wenshu, and Russell Smyth. 2011. "What Keeps China's Migrant Workers Going? Expectations and Happiness among China's Floating Population." *Journal of the Asia Pacific Economy* 16: 163–182.

Gong, Jin. 2013. "'Quality' of Gambling: Women Participating in Underground Lottery of Rural China." Accessed March 8. http://youthfieldcamp.org/2012-report-gong-j/

Guo, Yixian. 2005. *The Impacts of Illegal Lottey on the Financial Sector.* Accessed August 18, 2012. http://www.financialnews.com.cn/fz/200501310070.htm

Hansen, Ann. 1995. "The Tax Incidence of the Colorado State Lottery Instant Game." *Public Finance Review* 23: 385–398.

Harrell, Clyde Steva. 1974. *When a Ghost Becomes a God.* Stanford, CA: Stanford University Press.

Heitmueller, Axel. 2005. "Unemployment Benefits, Risk Aversion, and Migration Incentives." *Journal of Population Economics* 18: 93–112.

Herring, Mary, and Timothy Bledsoe. 1994. "A Model of Lottery Participation." *Policy Studies Journal* 22: 245–257.

Hing, Nerilee, and Helen Breen. 2001. "Profiling Lady Luck: An Empirical Study of Gambling and Problem Gambling Amongst Female Club Members." *Journal of Gambling Studies* 17: 47–69.

Hodge, David R., Kathleen Andereck, and Harry Montoya. 2007. "The Protective Influence of Spiritual-Religious Lifestyle Profiles on Tobacco Use, Alcohol Use, and Gambling." *Social Work Research* 31: 211–219.

Hodgins, D. C., D. P. Schopflocher, C. R. Martin, N. el-Guebaly, D. M. Casey, S. R. Currie, G. J. Smith, and R. J. Williams. 2012. "Disordered Gambling among Higher-frequency Gamblers: Who is at Risk?" *Psychological Medicine* 42: 2433–2444.

Hoffmann, John P. 2000. "Religion and Problem Gambling in the US." *Review of Religious Research* 41: 488–509.

Human Rights in China. 2002. *Institutional Exclusion: The Tenuous Legal Status of Internal Migrants in China's Major Cities*. London: Human Rights in China.

Humphreys, Brad R., Yang Seung Lee, and Brian P. Soebbing. 2011. "Modeling Consumers' Participation in Gambling Markets and Frequency of Gambling." *Journal of Gambling Business and Economics* 5: 1–22.

Hybels, Judith H. 1979. "The Impact of Legalization on Illegal Gambling Participation." *Journal of Social Issues* 35: 27–35.

Jackson, Raymond. 1994. "Demand for Lottery Products in Massachusetts." *Journal of Consumer Affairs* 28: 313–325.

Jaeger, David A., Thomas Dohmen, Armin Falk, David Huffman, Uwe Sunde, and Holger Bonin. 2010. "Direct Evidence on Risk, Attitudes and Migration." *Review of Economics and Statistics* 92: 684–689.

Jianakoplos, Nancy Amnon, and Alexandra Bernasek. 1998. "Are Women More Risk Averse?" *Economic Inquiry* 36: 620–630.

Joukhador, Jackie, Alex Blaszczynski, and Fiona Maccallum. 2004. "Superstitious Beliefs in Gambling among Problem and Non-problem Gamblers: Preliminary Data." *Journal of Gambling Studies* 20: 171–180.

Knight, John, and Ramani Gunatilaka. 2010. "Great Expectations? The Subjective Well-being of Rural–Urban Migrants in China." *World Development* 38: 113–124.

Lam, Desmond. 2006. "The Influence of Religiosity on Gambling Participation." *Journal of Gambling Studies* 22: 305–320.

Lee, Yu-Kang, and Chun-Tuan Chang. 2008. "A Social Landslide: Social Inequalities of Lottery Advertising in Taiwan." *Social Behavior and Personality: An International Journal* 36: 1423–1437.

Li, Xu. 2005. "The 'Red Ball' Affair." Accessed August 18, 2012. http://lixu1984.blogchina.com/81600.html

Li, Hai, Luke Lunhua Mao, James J. Zhang, Wu Yin, Anmin Li, and Jing Chen. 2012a. "Dimensions of Problem Gambling Behavior Associated with Purchasing Sports Lottery." *Journal of Gambling Studies* 28: 47–68.

Li, L., M. Morrow, and M. Kermode. 2007. "Vulnerable but Feeling Safe: HIV Risk Among Male Rural-to-Urban Migrants in China." *AIDS Care* 19: 1288–1295.

Li, Hai, James J. Zhang, Luke Lunhua Mao, and Sophia D. Min. 2012b. "Assessing Corporate Social Responsibility in China's Sports Lottery Administration and Its Influence on Consumption Behavior." *Journal of Gambling Studies* 28: 515–540.

Liang, Heng. 2004. "The Illegal Lottery Took 1.3 Billion Yuan a Year from Peasants in Guangdong Province." Accessed April 4, 2013. http://www.china.com.cn/chinese/2004/Aug/646803.htm

Liang, Huiru. 2009. "De Liuhecai Yu Mishi De Fa Shiyong." [The Prevalance of Illegal Lottery and Issues in Law Enforcement.] *Yanjiusheng faxue* [Graduate Law Review] 24: 1–12.

Liu, Xueting. 2011. "Speculation as Salvation: The Politics of Possibilities in Southeastern China's Underground Lotteries." Paper presented at the the American Anthropological Association annual meeting, November 16–20.

Loo, Jasmine M. Y., Namrata Raylu, and Tian Po S. Oei. 2008. "Gambling among the Chinese: A Comprehensive Review." *Clinical Psychology Review* 28: 1152–1166.

McCormack, Abby, Gillian W. Shorter, and Mark D. Griffiths. 2014. "An Empirical Study of Gender Differences in Online Gambling." *Journal of Gambling Studies* 30: 71–88.

McDaniel, Stephen R., and Marvin Zuckerman. 2003. "The Relationship of Impulsive Sensation Seeking and Gender to Interest and Participation in Gambling Activities." *Personality and Individual Differences* 35: 1385–1400.

Mikesell, John L. 1991. "Lottery Expenditure in a Non-lottery State." *Journal of Gambling Studies* 7: 89–98.

Ministry of Finance. 2013. "Sales of Lotteries as of December 2012." Accessed March 10. http://zhs.mof.gov.cn/zhuantilanmu/caipiaoguanli/201301/t20130111_727675.html

Nielsen, Ingrid, and Russell Smyth. 2008. "Who Wants Safer Cities? Perceptions of Public Safety and Attitudes to Migrants among China's Urban Population." *International Review of Law and Economics* 28: 46–55.

Nielsen, Ingrid, Russell Smyth, and Qingguo Zhai. 2010. "Subjective Well-being of China's Off-Farm Migrants." *Journal of Happiness Studies* 11: 315–333.

Perez, Levi, and Brad R. Humphreys. 2011. "The Income Elasticity of Lottery: New Evidence from Micro Data." *Public Finance Review* 39: 551–570.

Perez, Levi, and Brad Humphreys. 2013. "The 'Who and Why' of Lottery: Empirical Highlights from the Seminal Economic Literature." *Journal of Economic Surveys* 27: 915–940.

Petry, Nancy M. 2005. *Pathological Gambling: Etiology, Comorbidity, and Treatment.* Washington, DC: American Psychological Association.

Phongphaichit, Phasuk, Sangsit Piriyarangsan, and Nualnoi Treerat. 1998. *Guns, Girls, Gambling, Ganja: Thailand's Illegal Economy and Public Policy.* Chiang Mai: Silkworm Books.

Price, Donald I., and E. Shawn Novak. 2000. "The Income Redistribution Effects of Texas State Lottery Games." *Public Finance Review* 28: 82–92.

Pun, Ngai. 1999. "Becoming Dagongmei (Working Girls): The Politics of Identity and Difference in Reform China." *The China Journal* 42: 1–18.

Rabinovitch, Simon. 2013. "China's Lottery Boom Sparks Social Fears." Accessed April 22. http://www.ft.com/cms/s/0/b1eec222-5bd6-11e2-bf31-00144feab49a.html

Rogers, Paul, and Paul Webley. 2001. "It Could Be Us!: Cognitive and Social Psychological Factors in UK National Lottery Play." *Applied Psychology* 50: 181–199.

Schneider, Friedrich. 2005. "Shadow Economies Around the World: What Do We Really Know?" *European Journal of Political Economy* 21: 598–642.

Schneider, Friedrich, and Dominik H. Enste. 2000. "Shadow Economies: Size, Causes, and Consequences." *Journal of Economic Literature* 38: 77–114.

Scott, Frank, and John Garen. 1994. "Probability of Purchase, Amount Of Purchase, and the Demographic Incidence of the Lottery Tax." *Journal of Public Economics* 54: 121–143.

Sellin, Thorsten. 1963. "Organized Crime: A Business Enterprise." *The ANNALS of the American Academy of Political and Social Science* 347: 12–19.

Sharpe, Louise. 2002. "A Reformulated Cognitive-behavioral Model of Problem Gambling: A Biopsychosocial Perspective." *Clinical Psychology Review* 22: 1–25.

Smith, Chris, and Ngai Pun. 2006. "The Dormitory Labor Regime in China as a Site for Control and Resistance." *International Journal of Human Resource Management* 17: 1456–1470.

Smyth, Russell, Xiaolei Qian, Ingrid Nielsen, and Ines Kaempfer. 2013. "Working Hours in Supply Chain Chinese and Thai Factories: Evidence from the Fair Labor Association's 'Soccer Project'." *British Journal of Industrial Relations* 51: 382–408.

Stark, Oded. 1991. *The Migration of Labor.* Oxford: Blackwell.

Stark, Oded, and David E. Bloom. 1985. "The New Economics of Labor Migration." *American Economic Review* 75: 173–178.

Steinmüller, Hans. 2011. "The Moving Boundaries of Social Heat: Gambling in Rural China." *Journal of the Royal Anthropological Institute* 17: 262–280.

Stranahan, Harriet A., and Mary O. Borg. 1998a. "Horizontal Equity Implications of the Lottery Tax." *National Tax Journal* 51: 71–82.

Stranahan, Harriet A., and Mary O. Borg. 1998b. "Separating the Decisions of Lottery Expenditures and Participation: a Truncated Tobit Approach." *Public Finance Review* 26: 99–117.

Sun, Wanning. 2012. "The Poetry of Spiritual Homelessness: A Creative Pratice of Coping with Industrial Alienation." In *Chinese Modernity and the Individual Psyche*, edited by Andrew Kipnis, 67–88. New York: Palgrave Macmillan.

Tan, Andrew K. G., Steven T. Yen, and Rodolfo M. Nayga, Jr. 2010. "Socio-demographic Determinants of Gambling Participation and Expenditures: Evidence from Malaysia." *International Journal of Consumer Studies* 34: 316–325.

Tang, Catherine So-kum, and Tian Po Oei. 2011. "Gambling Cognition and Subjective Well-being as Mediators between Perceived Stress and Problem Gambling: A Cross-cultural Study on White and Chinese Problem Gamblers." *Psychology of Addictive Behaviors* 25: 511–520.

Tse, Samson, Alex C. H. Yu, Fiona Rossen, and Chong-Wen Wang. 2010. "Examination of Chinese Gambling Problems through a Socio-Historical-Cultural Perspective." *The Scientific World Journal* 10: 1694–1704.

Tversky, Amos, and Deniel Kahneman. 1992. "Advances in Prospect Theory: Cumulative Representation of Uncertainty." *Journal of Risk and Uncertainty* 5: 297–323.

Venkatesh, Sudhir Alladi. 2006. *Off the Books: The Underground Economy of the Urban Poor.* Cambridge, MA: Harvard University Press.

Wang, Wenfei Winnie, and C. Cindy Fan. 2012. "Migrant Workers' Integration in Urban China: Experiences in Employment, Social Adaptation, and Self-identity." *Eurasian Geography and Economics* 53: 731–749.

Welte, John W., Grace M. Barnes, William F. Wieczorek, Marie-Cecile Tidwell, and John Parker. 2001. "Gambling Participation in the US: Results from a National Survey." *Journal of Gambling Studies* 18: 313–337.

Wen, Ming. 2004. "Underground Lottery Has Emerged in Tianjin." Accessed September 1, 2012. http://news.sina.com.cn/s/2004-08-10/00193978872.shtml

Williams, Allan M., and Vladimir Balaz. 2012. "Migration, Risk and Uncertainty: Theoretical Perspectives." *Population, Space and Place* 18: 167–180.

Williams, Allan M., and Vladimir Balaz. 2014. "Mobility, Risk Tolerance and Competence to Manage Risks." *Journal of Risk Research* 17: 1061–1088.

Yen, Cheng-Fang, and Harry Yi-Hui Wu. 2013. "Gambling in Taiwan: Problems, Research and Policy." *Addiction* 108: 463–467.

Yoong, David, Tan Hooi Koon, and Ng Choung Min. 2013. "This is not Gambling but Gaming: Methods of Promoting a Lottery Gaming Company in a Malaysian Daily." *Discourse and Society* 24: 229–247.

Zeng, Weimin. 2004. *Dixia Liuhecai Jiemi* [Revealing the Secrets of Underground Mark Six]. Beijing: Minzu chubanshe.

Zeng, Zhonglu, and Dongmei Zhang. 2007. "A Profile of Lottery Players in Guangzhou, China." *International Gambling Studies* 7: 265–280.

Zhang, Yawei. 2007. "Guangnian, Xueli, Huanjin He Bocai Xingwei." [Attitude, Qualifcation, Environment and Gambling Behavior.] *Shijie jingji* [Journal of World Economy] 6: 48–55.

Zhang, Yawei, and Meiqin Zheng. 2006. "Hainan Bocai Canyulv Shizheng Fenxi." [An Empirical Analysis of Lottery Participation in Hainan.] *JIngji wenti tansuo* [Inquiry into economic issues] 4: 1–9.

Zhao, Yaohui. 1999. "Leaving the Countryside: Rural to Urban Migration Decisions in China." *American Economic Review* 89: 281–286.

Zhuang, Pinghui. 2013. *Migrant Workers Feel Like Outsiders in Mainland Cities, Says Survey.* March 3. Accessed April 29. http://www.scmp.com/news/china/article/1170903/migrant-workers-feel-outsiders-mainland-cities-says-survey

The effects of residential patterns and Chengzhongcun housing on segregation in Shenzhen

Pu Hao

David C. Lam Institute for East-West Studies, Hong Kong Baptist University, Kowloon Tong, Kowloon, Hong Kong

As cities in China undergo growth and transformation, they continue to absorb migrants from both ends of the economic spectrum, giving rise to socially mixed cities. As this occurs, the cities experience an elevated level of residential segregation due to the emergence of new forms of enclave urbanism, such as gated communities and *chengzhongcun* (villages-in-the-city). Factors including historical legacy, land institutions, and property-led development have contributed to this divided residential pattern at the neighborhood level. However, at larger geographical scales, the degree of segregation depends on whether the provision of different housing types is systematically segregated among urban districts. This paper, using Shenzhen as a case study, examines the spatial logic of the divided pattern of the population by analyzing the distribution of both urban residents and housing provisions. The analysis explores segregation between the privileged *hukou* holders and underprivileged non-*hukou* migrants as well as the spatial separation of formal urban housing and *chengzhongcun*. As expected, non-*hukou* migrants are largely segregated from *hukou* holders due to their much-constrained choice of housing and the widespread availability of *chengzhongcun*. A rather low degree of segregation is manifest at the sub-district level. The pattern is somewhat more desirable, as it maintains a more spatially equitable setting that enables disadvantaged groups to reside within short distances of jobs and amenities. Nevertheless, urban renewal programs targeted at *chengzhongcun* are most likely to jeopardize such a pattern of housing, which may aggravate segregation at the larger geographical levels.

1. Introduction

Residential segregation refers to the spatial concentration of ethnic or socioeconomic groups, often resulting in a residential mosaic across urban spaces. Ghettos, immigrant enclaves, and gated communities are some examples of segregated urban spaces. Caused by a voluntary or involuntary separation of people (Marcuse 1997), residential segregation works against the goal of assimilation and the integration of different social groups. Moreover, as resources and amenities are unevenly distributed across urban space, the locations of places of residence have potentially important consequences for the life chances of residents and their progeny. Where segregation is high, it is believed to play a large role in reducing societal opportunities in labor market participation and in areas including politics, education, and culture (Musterd 2005), often leading to concentrated and enduring poverty.

Ethnic or racial segregation is the most prominent form of segregation in European and North American cities, where the concentration and isolation of ethnic minority groups have received substantial attention. The results of such segregation are regarded as major hurdles for achieving social equality between ethnic majority and minority groups and furthering the assimilation process for immigrants. In China, however, ethnic segregation is only present at the regional level because of the distribution of ethnic minorities in the Western region and the Han dominance in most coastal and central cities (Li and Wu 2008). The intra-urban residential pattern is mainly manifest as socioeconomic segregation (Wu and Li 2005). While accommodating the new wealthy class, Chinese cities have also become home to a growing number of poor households, which are disproportionately rural migrants. The dramatic market transition has led to a widening wealth gap between urban residents (Li 2012) and an urban space that is increasingly divided (Madrazo and van Kempen 2012).

Since the mid-1990s, scholarly attention has been placed on the spatial structure, social space, and residential differentiation of Chinese cities. Lo (1994) examined the socialist city structure and found that residential differentiation is based on occupation types, which are related to the land use in the city. Yeh, Xu, and Hu (1995) found that the location of employment determines residential location, which is different from the social spatial structure of Western cities. Ma and Xiang (1998) examined the growing presence of migrants and diverse social groups who produce a new urban mosaic that did not exist in the pre-reform era. Huang (2004) analyzed the variety of housing types and identified increasing housing inequality and emerging residential segregation. More recently, Li and Wu (2008) found that residential segregation in Chinese cities occurs through differentiated housing types, which stem from pre-existing institutional privileges rather than household preference or life stages. Resulting from a legacy of state-led urban development and socialist housing provisions, urban poverty, which exists at the neighborhood level, is associated with certain housing types, such as dilapidated inner-city neighborhoods, declining work unit housing, and *chengzhongcun* (literately 'villages encircled by the city') (Liu and Wu 2006; He et al. 2010).[1] Most of the massive numbers of rural migrants are excluded from obtaining subsidized housing and can only afford low-rent rooms in locales like *chengzhongcun* (Song, Zenou, and Ding 2008). Such migrant enclaves, therefore, segregate rural migrants from privileged urban locals, the result of which has extensive social consequences.

It is thus hypothesized that the spatial distribution of different housing types might determine the residential pattern of different social groups, creating concentrations of privileged or under-privileged classes. What is the relationship between the housing supply of a city and the spatial division of its population? To what extent does the provision of cheap housing choices keep migrants confined to certain urban areas? And how do alterations in the quantity or spatial distribution of such provisions re-shape the residential pattern? These are pertinent questions crucial to gaining a better understanding of the causes and implications of residential segregation in urban China.

This paper examines the residential segregation of migrants and permanent local citizens with an emphasis on the provision of *chengzhongcun* housing in Shenzhen, one of the youngest metropolises in China. In 2011, of the 14 million residents in Shenzhen, 95 percent were born elsewhere, and only 2.3 million have a local urban *hukou* (Bach 2011). The rest of the population consists of 6.4 million migrants registered with a temporary residence permit (TRP) and more than 5 million totally unregistered migrants

(China Daily 2009). In this study, residents who possess a local urban *hukou* are identified as permanent local citizens. Those who lack a local urban *hukou* are considered 'migrants.' *Hukou* status does not imply a residence status or settlement intention; however, it stands as a significant barrier to urban rights and resources, and categorically discriminates against non-*hukou* residents. Judging from its short history and demographic profile of being overrepresented by non-*hukou* migrants, Shenzhen is far from a typical Chinese city. In the past, Shenzhen's territory consisted of fishing and farming villages as well as a small old town near the border with Hong Kong. In 1979, the whole territory had only 314,000 people (Shenzhen Statistics Bureau 2010). Now with its much larger population, Shenzhen has a very small proportion of genuine native residents (i.e. indigenous villagers and their descendants). The stratification of the citizens has little to do with whether they are natives or migrants. It is thus reasonable to study the differences between the privileged *hukou* population and the underprivileged non-*hukou* population. On the other hand, due to the lack of a historical city core and the almost-absence of pre-reform socialist housing, it is justifiable to study the dichotomy between the formal housing market[2] and the *chengzhongcun* market, which is considered informal and inferior (Zhang, Zhao, and Tian 2003; Wu, Zhang, and Webster 2013). Shenzhen, thus, provides a rather unique example of a contemporary Chinese city to use to study the prominent divisions in population and housing, although *chengzhongcun* prevail in almost all big cities.

This empirical analysis measures and compares the segregation patterns of urban residents and housing provisions over a large metropolitan area. To address the inconsistency of segregation measures due to the scale effect – one of the most intractable methodological problems in segregation research – multiscale analysis is employed. Multiscale analysis enables a more comprehensive understanding of the implications of residential segregation at different geographical and administrative levels. Dedicated local segregation indices are determined using high-resolution demographic and building data covering the entire city to examine the patterns of segregation in the population and housing at different scales, and to suggest reasons and implications for these patterns.

The next section discusses the socially divided pattern of Chinese cities. Section 3 introduces the methodology that allows for measuring multiscale segregation. Section 4 offers a brief discussion about the study area and data, and Section 5 measures and explains the segregation patterns in the population and housing types in Shenzhen. The final section provides policy implications and recommends future research topics.

2. Segregation with Chinese characteristics

Increased spatial mobility of capital and labor gives rise to greater segmentation of the urban space based on ethnicity, occupation, income, and other socioeconomic attributes. The processes of social differentiation inevitably result in residential segregation – the uneven distribution of social groups across space. Residential segregation refers to over- and under-representations of population groups in relation to the total population of an area (van Kempen and Özüekren 1998). It denotes socio-spatial inequalities, as groups living in different places have unequal access to resources and opportunities (Madrazo and van Kempen 2012). It also impedes interactions of social groups and the establishment of social relations.

Residential segregation has a long historical standing in Chinese cities. In imperial China, social mobility was limited. Residential separation between the ruling class and the masses was maintained through physical walls and building codes (Chang 1970). Both class and ethnic segregation existed, and were characterized by occupational homogeneity (such as the official elite) and personal wealth heterogeneity (Belsky 2000). However, even though Chinese cities in the imperial era had visible walls, their 'soft edges' allowed free urban–rural interaction and urban expansion (Lin 2007). Beginning in 1949, cities under Mao had the visible walls removed, but these were replaced by 'invisible walls' that controlled population mobility and separated the cities from the rural society (Chan 1994). The key policy instrument was the rigid household registration (*hukou*) system of rural–urban separation and differential treatment, which has underpinned China's social, economic, and political structures ever since (Chan 2010a). As Chan (2009, 215) stated, 'Serving as much more than a migration control mechanism, the *hukou* system was a mechanism for organizing labor in pursuit of Big Push[3] forced industrialization during most of the first three decades of the People's Republic. By immobilizing the peasantry and putting them under close surveillance, the state was able to orchestrate extractions from the agriculturalists to support the paramount goal of crash industrialization.'

With the rural population excluded from urban jobs, access to housing in cities was intimately tied to urban *hukou* status. Meanwhile, the *danwei* (work unit) system was introduced and the city was arranged into uniformly self-contained units where the urbanites lived and worked, and migrants were non-existent. Under the *danwei* system, where people lived and what amenities they had depended on their workplaces rather than their socioeconomic attributes (Logan, Bian, and Bian 1999). Moreover, the uniform apartment buildings, which provided housing to employees, housed people ranging from high-ranking officials to ordinary workers; therefore, the residential pattern in socialist Chinese cities became characterized by relatively homogeneous work-unit compounds. As housing was closely tied to workplaces, social spaces in the pre-reform era were mainly built upon different land uses rather than social stratification (Yeh, Xu, and Hu 1995). For instance, the living quarters of a state-owned enterprise housed the workers and managers; while the housing compound of a governmental unit accommodated both the cadres and their drivers.

The 1998 urban housing reform in China significantly changed cities' residential profiles. While it increased the overall supply and consumption of housing, it also dramatically worsened housing inequality and residential segregation (Huang 2004). With the privatization of public sector housing and an emphasis on commodity housing development, the absence of a local *hukou* has become less of an important constraint on housing consumption (Li 2012). However, *hukou* status still substantially dictates social and income classes (Chan 2009). Moreover, soaring housing prices exclude most migrants from accessing the newly established commodity housing market (Wang 2004), because more affordable housing options are only accessible to households with a local urban *hukou,* and migrants, regardless of their income level, do not qualify for such options. At the same time, market remuneration has become more important for determining one's socioeconomic status, and hence, housing consumption level (He et al. 2010). Because both the production and the consumption of housing has become partially market based (Li 2012), a relationship between social groups and housing types has been forged. People of different socioeconomic statuses are therefore increasingly divided through the designation and separation of housing types – from upscale, gated communities to low-income neighborhoods (Huang 2004; Liu and Wu 2006; Li, Zhu, and Li 2012). Residents of different housing types also exhibit varying space-time experiences in their daily lives, forming segregated spaces apart from housing for activities (Wang, Li, and Chai 2012).

In both Western and Chinese contexts, research has found that housing tenure is strongly correlated with segregation (Giffinger 1998; van Kempen and van Weesep 1998; Li and Wu 2008). The supply and location of owner-occupied dwellings, private rentals, and social rentals often lead to the spatial concentration of specific social groups through constraining people's housing choices. For instance, in Ireland, segregation arose when immigrants' housing options were limited to one particular form of housing tenure or location while access to other options of tenure and locations were open to natives (Vang 2010). In Britain, limited access to social housing, low rates of home-ownership, and the overrepresentation of the older housing stock in the private rental sector led to the concentration of immigrants in inner city neighborhoods (Peach 1998). Since the reform, increasingly complex tenure forms largely characterize the housing market in Chinese cities (Li and Yi 2007), where the provision of different housing types provides a key to understanding different tenures and the segregated urban population.

Among all housing types in urban China, *chengzhongcun* is probably the most inferior. It provides cheap housing, which is sometimes slum-like. As a consequence of China's urban–rural dual structure, the development and administration of *chengzhongcun* are largely autonomous. Houses in *chengzhongcun* neighborhoods are self-built by local villagers, and the collectives of the villagers are responsible for the provision of public goods including infrastructure, sanitation, policing, social welfare, and education (Po 2012). *Chengzhongcun* neighborhoods, though encircled by the city and accommo-dated by mostly urban dwellers, are not even considered urban spaces by many scholars (Li 2004; Lan 2005), government authorities (Shenzhen Public Security Bureau 2005; Shenzhen Urban Planning Bureau 2005), and the media (*People's Daily's* 2004; *Southern Metropolis Daily* 2011). Instead, they are often viewed in sharp contrast with formal housing neighborhoods as backward places and oddities associated with poor residents and inferior living environments (Wang, Wang, and Wu 2009; Wu 2009; Bach 2010). However, since *chengzhongcun* are physically and socioeconomically interwoven with surrounding urban areas, they have become an integral component of the urban housing market and land use (Hao et al. 2012). *Chengzhongcun* also allow the indigenous villagers to earn their livelihoods as landlords (Zhang, Zhao, and Tian 2003), thereby simultane-ously solving the housing problem of rural migrants and the unemployment of landless farmers. The dichotomy between formal urban housing and *chengzhongcun* is therefore a prominent feature in urban China, to a great extent, reflecting its underlying dual social structure and is worth more analysis.

Equally prominent is the dichotomy between permanent local citizens and migrants without a local urban *hukou*. Non-*hukou* migrants have long been considered 'second-class citizens', and have been deprived of institutional access to urban welfare and pub-lic services (Chan 2009). Compared to *hukou* holders, non-*hukou* migrants typically have lower-paying jobs, poorer living conditions, less education, a lack of social security, and uncertainty in their long-term livelihood (Fu and Ren 2010; Wang and Wu 2010; Zhao and Howden-Chapman 2010). Urban housing for rural migrants is espe-cially constrained by land and housing institutions (Song, Zenou, and Ding 2008). In many cities, *chengzhongcun* provide the only affordable and accessible housing for rural migrants who are otherwise shunned by the urban housing market (Zhang 2005; Song, Zenou, and Ding 2008).

The two dichotomies have significant social consequences and have been studied extensively, but no research has explicitly unraveled the role of the former in shaping the residential profile of the latter. In other words, it is still unknown to what extent *chengzhongcun* segregate non-*hukou* migrants from the local *hukou* holders. To bridge this gap, the empirical analysis that follows provides a consistent depiction of the segregation patterns of population and housing provisions at multiple levels. The analysis explores segregation between privileged *hukou* holders and underprivileged non-*hukou* migrants as well as the spatial separation of formal urban housing and *chengzhongcun*. It further attempts to uncover the connection between the most prominent division in housing and the most prominent division in urban population – a topic with important implications for policy-making, as the central government has been calling for a 'new style' of urbanization focused on making cities fairer for migrants.

3. Measuring multiscale segregation

Though significant social implications exist with segregation, it is inherently a geographical problem (Brown and Chung 2006). Segregation examines whether different phenomena are more divided or mixed in space. To measure the levels of such spatial division and mix, segregation indices like the index of dissimilarity (ID) (Duncan and Duncan 1955) are the most effective and widely used methods. First employed to measure the segregation of the white and black population in American cities, the ID is defined as:

$$D = \frac{1}{2} \sum_i \left| \frac{w_i}{W} - \frac{b_i}{B} \right|$$

where w_i and b_i are black and white population counts in spatial unit i, and W and B are the total white and black population counts of the entire region. It ranges from 0 to 1, indicating no segregation to complete segregation, respectively. If all units have the same proportion in the two groups, then there is no segregation, and D is zero. If one group exclusively occupies each spatial unit, then D is one, indicating a completely segregated pattern.

Segregation measures like ID are highly sensitive to spatial scales. As the index D is purely a function of the homogeneity within a spatial unit, in general, the smaller the spatial unit, the more homogeneous the population mix and thus the higher the value of D. When data are aggregated to adjacent values, they are spatially 'smoothed,' and thus, less variation is preserved at the aggregated level. Therefore, when different levels of aggregation are available, values from segregation measures will be higher at the levels with smaller spatial units than at the levels with larger spatial units. This is referred to as the scale effect of the so-called modifiable spatial unit problem (Wong, Lasus, and Falk 1999; Wong 2004). To deal with the problem, multiscale analyses are recommended to provide more comprehensive insight into the spatial segregation pattern of population (Fotheringham 1989). In this regard, Geographic Information Systems can be employed to perform spatial analysis repeatedly on data with multiple scales to assess the scale effect (Wong 2003).

To examine and relate the segregation values obtained from different scale levels, Wong (2003) proposes a methodological framework to decompose segregation values derived from the dissimilarity index at the local level to pure local segregation and regional segregation. The decomposition accounts for the sources of segregation at different scale levels so that segregation values from different scales can be associated conceptually. This approach can provide a consistent accounting of segregation across multiple scales as long as the spatial units at different scale levels are nested in a hierarchical manner. Moreover, local segregation measures can be derived to reflect the degree of segregation attributable to the overall segregation level from local units at a given geographical level. This spatial decomposition framework is adopted in this study to develop a consistent depiction of the segregation patterns of population and housing provisions at multiple levels.

According to the decomposition framework by Wong (2003), the regional segregation (RD$_j$) of region Ω_j is calculated using:

$$RD_j = \left| \frac{\sum_{i \in \Omega_j} w_{ij}}{W} - \frac{\sum_{i \in \Omega_j} b_{ij}}{B} \right|$$

Meanwhile, the local segregation is introduced by the local unit level for region Ω_j. When the regional segregation is separated, the local segregation (LD$_j$) can be defined as:

$$LD_j = \sum_{i \in \Omega_j} \left| \frac{w_{ij}}{W} - \frac{b_{ij}}{B} \right| - \left| \frac{\sum_{i \in \Omega_j} w_{ij}}{W} - \frac{\sum_{i \in \Omega_j} b_{ij}}{B} \right|$$

Different from ordinary local segregation measures (such as the location quotient [LQ]), the local D (LD) index measures the contribution of segregation only from the local units, or the segregation at the local level conditioned by the regional level. It is relatively low if all local units within the same region have a similar composition for the two groups (i.e. local units have strong positive spatial autocorrelation). When the local D is mapped, it can indicate the spatial variation of segregation contributed purely by the local units. On the other hand, the regional D is the level of segregation that each region contributed to the segregation of the entire study area. Using both the local D and regional D allows us to explore the spatial variations of segregation at different scales and possibly from different sources. To supplement the new local measures, a conventional measure of the LQ is used to examine the uneven distribution of population and housing, which indicates a single-unit concentration and treats each spatial unit independently. In addition, LQ measures enable a comparison of the segregation pattern of Shenzhen with that of other cities, such as Shanghai (Li and Wu 2008) and Nanjing (Liu and Wu 2006).

Using high-resolution data on Shenzhen, the distribution values of population and housing provisions can be aggregated at multiple geographical scales. The regional and local measures of ID as well as LQ measures can be calculated and mapped to identify segregation patterns at multiple scales. The map from the regional D measure illustrates how each regional unit deviates from the population mix of the entire study area. Meanwhile, the map from the local D measure illustrates the variation of the population mix across local units within each regional unit. A high value of local D (LD) implies there is a large variation in the population mix within a regional unit, and a high level of segregation across local units contributes to the overall segregation level of the entire study area. On the other hand, a low LD indicates a rather homogeneous population mix within a regional unit, and a low level of segregation across local units contributes to the overall segregation level of the entire study area. Before applying these measures to the empirical data, a brief introduction to the study area and data are presented next.

4. Study area and data

Shenzhen is a young, migrant city established in 1979 as an experiment of China's open-door policy. The operation of a market economy in Shenzhen has enabled its astonishing urbanization and economic growth. An emphasis on export-oriented industries has turned the city into a major manufacturing center that continues to attract migrant workers, most of whom were denied a local urban *hukou*. Shenzhen's rapid economic growth drives an equally large area where the mass production of the built environment with a certain level of rurality persists along with the presence of pre-existing rural communities (Chung 2014). Spatial expansion of urban land has encroached its rural hinterland, creating 320 *chengzhongcun*. With urban development surrounding them, *chengzhongcun* are distributed throughout the city and have become an interwoven component of the urban economy and society (Hao, Sliuzas, and Geertman 2011). Despite the important role of *chengzhongcun*, the government maintains a negative view of them, claiming that they are associated with physical and social problems, and that their existence suppresses the value of the land where they are located and in their neighboring areas. Consequently, policies have been implemented aiming to solve the '*chengzhongcun* problem' through a wholesale demolition and redevelopment into formal urban neighborhoods. In 2005, Shenzhen introduced the Master Plan of *Chengzhongcun* Redevelopment 2005–2010 (Shenzhen Urban Planning Bureau 2005), which was the first official move to demonstrate the determination of the government to halt the further development of *chengzhongcun* (Chung 2009). The move also made Shenzhen the first Chinese city that adopted a set of systematic policies to implement a 'comprehensive', all-out full-scale *chengzhongcun* redevelopment.

Shenzhen used to be composed of two distinctive administrative divisions: a Special Economic Zone (SEZ), functioning as the city center for tertiary development; and a non-SEZ area (which was integrated into Shenzhen in 1993), featured by extensive industrial development. In both regions, former collective farms were replaced with built-up urban landscapes, and village settlements that had been left untouched became *chengzhongcun*, one of the most conspicuous by-products of Shenzhen's urban explosion. In a state of quasi-urban jurisdiction where rural collective land ownership remains, *chengzhongcun* enjoyed a high level of autonomy, allowing the indigenous villagers to build tenement houses for migrant tenants. The 320 *chengzhongcun*, in the form of thousands of settlement fragments, are distributed over both the SEZ and non-SEZ areas (Figure 1), but *chengzhongcun* exhibit different characteristics in terms of their phase of development and local contexts. They also face different policy environments, which exert different influences on their development trajectories (Hao et al. 2013a).

In 2009, *chengzhongcun* provided a housing market of a total floor space of 173 million m^2, equivalent to 42 percent of the total floor space of the entire housing sector, or 23 percent of the total floor space of all buildings in Shenzhen (Hao et al. 2013b). *Chengzhongcun* are thought to accommodate over 7 million people, the vast majority of whom are rural migrants (Zacharias and Tang 2010). The physical growth of *chengzhongcun* during 1999–2009 was striking, contributing to an increase in floor space of 105 million m^2. In contrast, the development of commodity housing yielded 58 million m^2 of floor space in the same period (Shenzhen Statistics Bureau 2010). The scale and speed of development of *chengzhongcun* in Shenzhen were likely the most extreme of any city in China (Hao, Sliuzas, and Geertman 2011). This was due to some unique features of the city: (1) its proximity to Hong Kong determined its designation

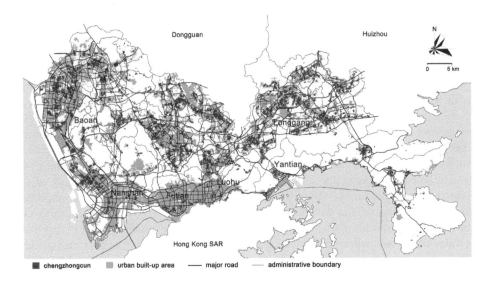

Figure 1. The distribution of *chengzhongcun* in Shenzhen (2009).
Source: Urban Planning and Land Resources Commission of Shenzhen Municipality.

as a SEZ and the practices of an export-oriented economy, and (2) the rapid development of the city from scratch, which relied on negotiating with local farmers to expropriate land, enabling farmers to keep their housing land and granting farmers permission to develop their low-density village settlements into high-density neighbor-hoods (Zhang, Zhao, and Tian 2003). The city's economic success is also attributed to the unprecedented influx of migrant laborers. The migrants generate an increasing demand for low-cost housing, which is readily available and continuously being developed in *chengzhongcun* (Hao et al. 2013a).

The development of *chengzhongcun* has been propelled by the inflow of migrants. From 1979 to 2008, the annual average growth rate of the floating population was 33.5 percent, much higher than that of the local *hukou* holders, which was 7.1 percent (Shenzhen Statistics Bureau 2009). Consequently, the proportion of the floating population in Shenzhen has been dramatically rising. In 2005, 4.69 million non-*hukou* migrants lived in *chengzhongcun*, 14 times the number of indigenous villagers, which was about 0.33 million (Shenzhen Public Security Bureau 2005). These figures illustrate the importance of *chengzhongcun* both as a source of affordable housing for migrants and as a source of livelihood for indigenous villagers. The *chengzhongcun* also alleviate the burden on the government to provide housing and jobs for the landless peasants and lower the risk of associated social unrest (Hao, Sliuzas, and Geertman 2011). Nevertheless, the spatial division of *chengzhongcun* housing and ordinary living facilities created a highly segregated residential profile (Zhang 2005). The degree of such segregation at different geographical scales and the extent to which the segregation is a result of the presence and growth of *chengzhongcun* will be examined using 2009 Shenzhen data.

In 2004, all *chengzhongcun* in Shenzhen were converted from rural to urban jurisdictions as residents' committees. Consequently, residents' committees that comprise *chengzhongcun* and residents' committees of non-*chengzhongcun* neighborhoods share the same level of administrative division, which is under the upper division of the sub-district. The spatial units of *chengzhongcun* and other residents' committees are mutually exclusive geographic entities, and the two types of spatial units can thus be clearly discriminated. Although Shenzhen's residents' committees vary in size, in general the areal size of a committee is similar to that of nearby committees, and the size is proportional to that of its nesting unit (see Figures 3–6). In other words, a district of larger areal size consists of sub-districts of a larger areal size, each of which again consists of residents' committees of a larger areal size. This proportional size distribution between nested and nesting units automatically controls the effect of the size variance of spatial units on segregation measures, which further justifies the application of multiscale analysis to the case of Shenzhen.

Among the three regular sources of population information in China – the Civil Affairs Bureau, the Public Security Bureau, and the census – the census typically has the most representative demographic data for a large territorial unit, such as a municipality, a province, or the entire country. However, for Shenzhen, the 2010 Census includes 10.4 million people, far less than the generally accepted and quoted population count of 14–15 million (with a large presence of highly mobile migrant workers) (Zacharias and Tang 2010; Bach 2011; Xinhua News 2011; The Guardian 2014). While the actual number of people in Shenzhen remains in dispute, the most suitable data for this research come from the Household Registration Database (HRD) held by the Shenzhen Public Security Bureau. The HRD keeps a record of *hukou* residents and non-*hukou* residents who have obtained a TRP (*zanzhu zheng*). Holding a local *hukou* or a TRP is mandatory for accessing the formal job and housing markets in the city. Migrants who have not obtained proper paperwork such as the TRP, if living in informal housing such as *chengzhongcun* and working in the informal sector, are usually not covered by the official databases and the census. However, local police stations at the grassroots-level track the presence of households and residents (including those unregistered ones) in each community and record the population information in the HRD. Consequently, the 2009 HRD data of 14.8 million resident records is the most comprehensive of Shenzhen's population profile, and the database, instead of the census, is used by most government agencies for policy-making and planning.[4]

The data on housing types and their capacity and locations are provided by the Shenzhen Municipal Building Database. The 2009 building data cover physical and spatial information of all 615,702 buildings in Shenzhen, including all of *chengzhongcun* houses and formal urban houses.

The following analysis primarily uses these two databases to determine various segregation measures and to compare the spatial segregation of non-*hukou* residents and local *hukou* residents, as well as of *chengzhongcun* housing and formal urban housing.

5. Segregation and housing provision in Shenzhen

Of Shenzhen's 14.8 million people, only 2.3 million people are local *hukou* holders (15.2 percent); the majority of the city's 12.6 million migrants (84.8 percent) do not have a local *hukou*. In all six districts, the proportion of migrants is considerably higher than the proportion of local *hukou* holders (Table 1). It is also evident that the four SEZ districts – Futian, Luohu, Yantian, and Nanshan – have much higher percentages of local *hukou* holders (22–31 percent) than the non-SEZ districts (7–8 percent) (Shenzhen Public Security Bureau 2009). This is primarily due to the structural variance in the development of different sectors and urban functions among the six districts, which have concentrations of different social groups. The SEZ districts are home to company headquarters, government offices, and high-end service sectors, all of which provide better

Table 1.   The composition of population and housing types in Shenzhen (2009).

	Population (million)					Housing (million m²)				
	Non-*hukou*		*Hukou*		Total	*Chengzhongcun*		Non-*chengzhongcun*		Total
Futian	1.43	68.8%	0.65	31.3%	2.08	7.90	16.0%	41.63	84.0%	49.53
Luohu	1.38	74.0%	0.48	26.0%	1.86	6.79	18.1%	30.72	81.9%	37.51
Yantian	0.25	77.3%	0.07	22.7%	0.32	1.07	19.7%	4.36	80.3%	5.43
Nanshan	1.16	73.9%	0.41	26.1%	1.57	9.21	20.6%	35.56	79.4%	44.77
Baoan	5.21	93.5%	0.36	6.5%	5.57	73.85	47.6%	81.40	52.4%	155.25
Longgang	3.15	91.8%	0.28	8.2%	3.43	59.33	49.0%	61.75	51.0%	121.08
Total	12.57	84.8%	2.26	15.2%	14.83	158.16	38.2%	255.41	61.8%	413.57

pay and more secure jobs. In the non-SEZ districts, the prevalence of large-scale industries and lower end services is a magnet for non-*hukou* migrants, because on the one hand, non-*hukou* migrants are often denied more privileged jobs and on the other hand, without a 'decent' job, obtaining an urban *hukou* is extremely difficult (Chan 2009; Zhang, Zhu, and Nyland 2012).

In Shenzhen, *chengzhongcun* provide 38 percent of the total residential floor space, while formal urban housing, including commodity housing, public housing, and workers' dormitories, provide the other 62 percent (Table 1). Nevertheless, *chengzhongcun* house half of the total population (Zacharias and Tang 2010), reflecting a substantial contribution to the city. The data also indicate that *chengzhongcun* housing is characterized by a significantly higher living density than formal urban housing. Given that the majority of the migrant population is excluded from the formal housing market in Shenzhen (Song, Zenou, and Ding 2008), *chengzhongcun* housing offers the most important housing option for disadvantaged groups. Because of the involuntary bond between the migrant population (the disadvantaged in the population dichotomy) and *chengzhongcun* housing (the inferior in the housing dichotomy) discussed in Section 2,

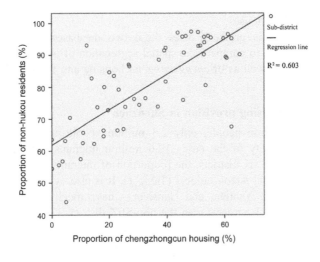

Figure 2.   Correlation of migrants with *chengzhongcun* housing.

the proportion of non-*hukou* residents and the proportion of *chengzhongcun* housing are highly correlated (Figure 2).

Across the six districts of Shenzhen, neither *chengzhongcun* housing nor the formal urban housing is evenly distributed. *Chengzhongcun* housing is mostly distributed in Baoan and Longgang, where this particular housing market supplies almost half of the total housing space. The four SEZ districts still maintain a large stock of *chengzhongcun* housing, with percentages of total residential floor space that range from 16 to 21 percent. In the SEZ districts, the smaller number of *chengzhongcun* and the smaller areal size of each *chengzhongcun* determine a lower share in the total housing provision than their counterparts in the non-SEZ districts; however, the building and population densities of those centrally located *chengzhongcun* are the highest in Shenzhen (Hao et al. 2013a).

Figure 3 shows the LQ for migrants in Shenzhen at the level of the residents' committee. The spatial division is manifest: the city is characterized by an over-representation of migrants in the non-SEZ districts and an under-representation of migrants in the SEZ districts. Most residents' committees in Futian, Luohu, and Nanshan are inhabited by an above-average level of local *hukou* holders, while pockets of migrant concentrations also exist in these districts. In contrast, residents' committees in Baoan and Longgang are predominantly inhabited by an above-average number of migrants. In the non-SEZ districts, low LQ units form two large clusters within the mass of high LQ units. One cluster located in northern Baoan is Guangming, the only substantial farm area the city preserves for recreation, urban agriculture, and light agricultural processing industries. The other cluster, located in the Dapeng Peninsular at the east end of Longgang, consists mostly of designated ecological protection zones. These two clusters are much less urbanized compared to other parts of the city, and thus they have attracted fewer migrants, who account for a significantly lower proportion than the average proportion in the city.

As *chengzhongcun* provide a major housing market for migrants, the LQ for *chengzhongcun* housing demonstrates a pattern (Figure 4) similar to the one for the migrant population. The city center is homogeneous with an over-representation of

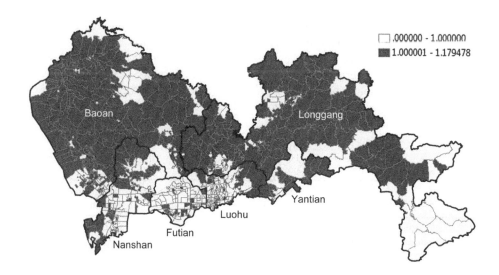

Figure 3.   Location quotient for non-*hukou* residents at the level of the residents' committee.

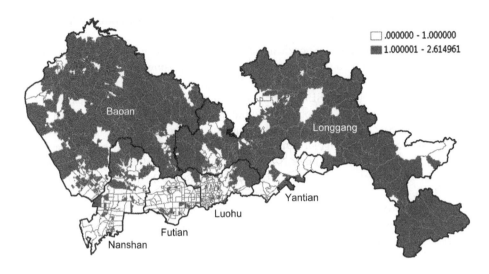

Figure 4.   Location quotient for *chengzhongcun* housing at the level of the residents' committee.

non-village housing. As *chengzhongcun* primarily form a number of residents' committees (the boundaries of *chengzhongcun* coincide with the boundaries of residents' committees), these committees have become pockets of high LQ units in the SEZ districts where formal urban housing is dominant (see Table 1). In the non-SEZ districts, residents' committees are mostly characterized by an over-representation of *chengzhongcun* housing. In Guangming and Dapeng, the low level of urbanization also leads to a lack of commodity housing development, and therefore an over-representation of *chengzhongcun* housing is present in the two regions.

The uneven distribution of population and housing across space at three levels – district, sub-district, and residents' committee – is captured by the ID (Table 2). Residential segregation at larger levels is low, because while the uneven distribution of the urban population is significant at smaller scales, when the data are aggregated to adjacent values, less variation is preserved at the aggregate level. However, even at the level of the district, segregation between migrants and local *hukou* holders is considerably high, which is largely due to the over-representation of local *hukou* holders in the SEZ districts and the over-representation of migrants in the non-SEZ districts. There is relatively little variation between the SEZ districts or between the non-SEZ districts. This pattern is a direct result of the functional difference between the city's SEZ and non-SEZ areas.

Table 2.   Segregation of migrants and housing types at different levels.

	Migrants vs. *hukou* holders	*Chengzhongcun* housing vs. other housing
Index of dissimilarity (district)	0.379	0.282
Index of dissimilarity (sub-district)	0.418	0.315
Index of dissimilarity (residents' committee)	0.593	0.538

If all *chengzhongcun* in Shenzhen are considered a sub-market, there is large variation within this sub-market that spreads across the six districts. Population density in the SEZ districts, where the ratio of landlords and tenants is about 1:30, is higher than in the non-SEZ districts, where the ratio is about 1:10. As urban development is diverse and migrants with different socioeconomic statuses are unevenly distributed, the social structures of *chengzhongcun* residents vary across districts. Similar to formal urban housing, *chengzhongcun* also function as diverse housing markets. The socioeconomic status of the residents of *chengzhongcun* is often linked to the types of jobs available in the respective district (Hao et al. 2013b). For instance, in the SEZ, significant proportions of *chengzhongcun* tenants are office and tertiary sector employees. Outside the SEZ, the majority of *chengzhongcun* tenants are industrial workers and employees in small and/or informal businesses. Because the population composition of migrants and local *hukou* holders is much more unbalanced than the housing-type composition of *chengzhongcun* and formal housing, the ID for population is higher than the ID for housing provisions (Table 2).

At the level of the sub-district, the ID between migrants and local *hukou* holders is 0.418, and the ID between village housing and other housing is 0.315. Judging from the small increase in ID from the district level to the sub-district level (i.e. from 0.379 to 0.418 for population, and from 0.282 to 0.315 for housing), segregation at the sub-district level mostly stems from the segregation at the district level, especially from the difference between the SEZ and non-SEZ districts. At the sub-district level, since *chengzhongcun* are distributed across almost all units, the added degree of segregation from the district to sub-district level (indicated by the increase in the ID values) is limited.

At the level of the residents' committee, segregation is extremely high for both population and housing provisions. While a residents' committee has an average population of about 23,000, the ID indicating segregation between migrants and local *hukou* holders is almost 0.6. To put it in perspective, in Shanghai, a residents' committee has an average population of only 3000, and the ID between migrants and local *hukou* holders is lower than 0.3 (Li and Wu 2008). It is expected that the level of segregation in Shenzhen would be higher than 0.6 if an even smaller residential unit were examined. In accordance with the residential segregation, it is evident that *chengzhongcun* housing and formal urban housing are also highly segregated, as indicated by a very high ID of 0.538.

When the ID indices discussed above are decomposed, local measures of segregation pertaining to multiple geographical levels are computed according to the spatial decomposition framework of Wong (2003). These local segregation measures indicate the levels of segregation contributed by the local units and the regional units to the entire city. At the level of the district, Baoan has the highest level of RD (regional segregation) between migrants and local *hukou* holders, suggesting the over-representation of migrants in Baoan. The RD of Futian is moderately high due to the under-representation of migrants in Futian (Figure 5(a)). In terms of housing provisions, both Baoan and Longgang have the highest RD because of the higher proportion of *chengzhongcun* housing in the two districts (Figure 5(b)). The moderately high RD of Futian is a result of the under-representation of *chengzhongcun* housing in the city's central business district (see Table 1).

Though the maps of LD (local segregation) (Figure 5(c) and (d)) show patterns similar to those of the RD maps, they uncover the intra-district segregation of population and housing provisions, regardless of the segregation across districts. For

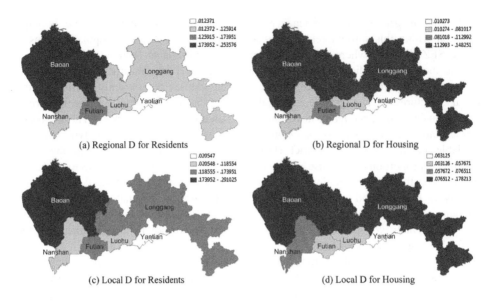

Figure 5.   Decomposed segregation measures at district level.

population data, the LD indicates that the level of segregation between migrants and local *hukou* holders is the highest in Baoan, followed by Longgang and Futian. Baoan has the most segregated districts at both the regional and local levels. While Baoan has the highest concentration of migrants, certain sub-districts in Baoan have become the most segregated migrant enclaves. Although the migrant population in Futian is under-represented, the relatively high local D indicates that the small proportion of migrants in Futian is highly segregated at the sub-district level. Luohu and Nanshan have relatively low levels of local segregation. In Yantian, the level of segregation of migrants and local *hukou* holders is the lowest.

The LD for housing provisions identifies that Baoan and Longgang have the highest level of local segregation between *chengzhongcun* housing and formal urban housing. Some sub-districts are concentrated with more *chengzhongcun* housing than other sub-districts. When *chengzhongcun* housing is more scattered in distribution, the local segregation becomes comparatively less prominent. The non-SEZ districts are the most segregated at both the regional and local levels. *Chengzhongcun* in certain sub-districts in Baoan and Longgang are the most segregated residential areas in Shenzhen.

At the level of the sub-district, Figure 6 shows the regional and local Ds for population and housing provisions, respectively. For both population and housing, RDs identify the highest valued sub-districts in both the center and peripheries of the city. However, similar to what was explained for the district level RDs, a high RD in the city center is actually the result of an under-representation of migrants and *chengzhongcun* housing. The high RD in the non-SEZ districts (Figure 6(a) and (b)), which is due to the concentration of migrants or *chengzhongcun* housing in certain sub-districts, explains the high local Ds shown in Figure 5(c) and (d).

From Figure 6(c) and (d), the segregation identified by the LD shows a completely different pattern. The comparison between Figure 6(a) and (c) indicates that units with the highest regional D exhibit a low local D. This demonstrates that when different spatial scales are under scrutiny, the degree of segregation may vary significantly. Certain

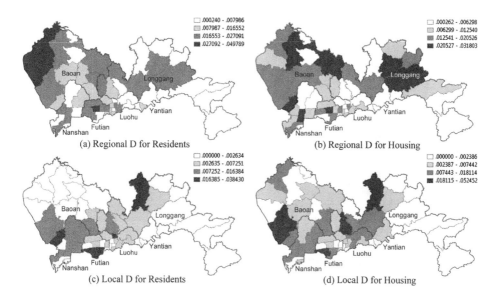

Figure 6.   Decomposed segregation measures at sub-district level.

sub-districts accommodate a larger proportion of migrants and thus exhibit higher segregation at the regional level, but if the migrants living in the sub-district are relatively more evenly distributed within the sub-districts, it will show low segregation at the local level. Moreover, in a migrant city like Shenzhen, urban residents also include migrants who are highly educated and have good-paying jobs. They do not possess a Shenzhen *hukou* because of institutional constraints or the preference for keeping their *hukou* affiliated with another place. The majority of migrants are characterized by lower-middle and low socioeconomic statuses, but all do not necessarily live in the *chengzhongcun*. Other types of low-cost housing, including workers' dormitories provided by the employer, also accommodate a considerable proportion of low-income migrants.

The measure of regional and local D indices illustrates how the segregation of migrants and local *hukou* holders is created at each geographical and administrative level. While segregation at the district level is already significant due to the concentration of local *hukou* holders in the SEZ, segregation at the sub-district and residents' committee levels adds another dimension to the problem. At the sub-district level, the non-SEZ districts manifest high degrees of residential segregation, probably because of the differentiated development of local economies. But at the residents' committee level, the non-SEZ districts generally have low residential segregation because of a highly even distribution across these small spatial and administrative units. In contrast, due to the demolition of numerous *chengzhongcun* in the city center, the SEZ districts suffer much greater residential segregation at the smaller scales. *Chengzhongcun* that remain in the center become pockets of migrant concentration and the most segregated enclaves (Figure 7).

Wholesale redevelopment programs have gradually been reducing the stock of *chengzhongcun* housing, converting *chengzhongcun* into formal urban neighborhoods. Simultaneously, most of the tenants cannot afford the much higher rents of local units in the formal housing market and have been expelled from those redevelopment areas.

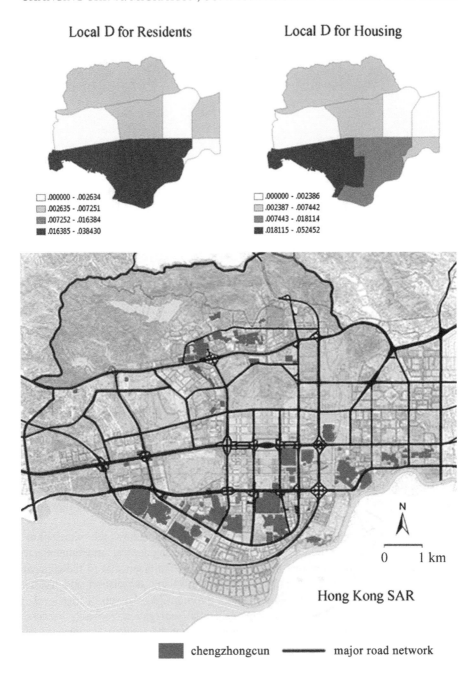

Figure 7.   Segregation patterns and *chengzhongcun* in Futian district.

The remaining *chengzhongcun* in the city center have become fewer and smaller. When the concentration patterns of migrants and *chengzhongcun* housing are examined, these villages are undoubtedly a source of segregation (Figure 7). However, these small pockets of accessible living locales in central locations are a haven for socially and economically disadvantaged groups as well as a springboard for the newcomers who strive for success and better lives in this increasingly exclusive city. Although *chengzhongcun* residents are segregated from the local *hukou* holders at an early stage of their life, being spatially well positioned in the city, some proportion of them will manage to move up the social ladder and seek better housing. With the market reigning supreme, their access to the city could be at risk if they cannot afford the rising housing costs. Redevelopment, which uproots some residents, further jeopardizes the chances for large segments of the urban poor to have a continuing claim to the city (Li 2012). As real estate developers have selectively redeveloped central urban areas because of the potential for profits, the displacement of *chengzhongcun* residents is likely to re-establish new segregated areas and increase segregation at higher levels.

6. Conclusions and discussion

This paper identifies the degrees of residential segregation attributable to different geographical levels. The analysis uncovers that residential segregation in Shenzhen is formed in a structure of systematic division at three administrative levels: the separation of the population between the SEZ and the non-SEZ due to their distinct modes of development; the separation of the population among sub-districts due to specific local economies and associated employment opportunities; and the separation of the population among residents' committees due to the availability of housing types. Though the entry–exit control of the SEZ was abolished in the late 1990s, the SEZ border still has a significant impact as an 'invisible wall.' Migrant laborers are allowed into the city, but they are 'used' as merely a means of production and have been confined to the labor-intensive sector of the city. By and large, non-*hukou* migrants are 'walled' in *chengzhong-cun*, which are still under de facto rural collective administration and in urban districts only recently expropriated from rural communities. In China's big cities, the pre-reform discrimination against underprivileged rural population continues to prevail.

Given their income levels and *hukou* status, most migrants are limited to residing in *chengzhongcun,* which are economically affordable housing types that are institutionally accessible in that you do not need a local *hukou* to rent a room. In a sense, the constellation of *chengzhongcun* has molded the influx of migrants into a largely predetermined shape, consisting of numerous pockets of cheap housing across the city The presence and physical growth of *chengzhongcun,* both of which are a legacy of the pre-existing rural communities, have had a substantial role in forming the current pattern of residential segregation. However, the role of *chengzhongcun* is two-fold. At the neighborhood level, *chengzhongcun* are undoubtedly a form of segregation rather than a locus for genuine integration, but at higher geographical and administrative levels, they allow the penetration of migrants into the capital-intensive and resource-rich sections of the city and enable a somewhat more even distribution of migrants across urban sections. In so doing, they reduce residential segregation at larger geographical levels. By allowing tenants to live in prime urban locations, *chengzhongcun* provides migrants closer proximity to employment and amenities, and thus better situates them to improve their life chances.

In view of the multiscale segregation pattern, Shenzhen's urban regeneration policy deserves a critical review. The policy has been established to demolish and redevelop all centrally located *chengzhongcun,* which are viewed as eyesores, poverty enclaves, and breeding grounds for social problems (Chung 2009; Hao, Sliuzas, and Geertman 2011). But simply redeveloping *chengzhongcun* cannot eliminate poverty or segregation.

Instead, wholesale redevelopment projects merely push poverty out of the central city and exacerbate segregation at larger geographical and administrative scales. While alternative local housing is unavailable, redevelopment programs would force displaced tenants to move to other more remote *chengzhongcun* and substantively increase commuting costs and the likelihood of unemployment. Moreover, as the policy primarily targets important urban segments and places the emphasis on redeveloping *chengzhongcun* located in major administrative and commercial areas and neighboring sites of major infrastructure projects (such as new metro lines), the current and future centers of urban resources are likely to exclude low-income migrants and deprive them of opportunities in the labor market and of access to public amenities.

Shenzhen's economic success is enabled by the unprecedented influx of migrant laborers, consisting of a disproportionately large population of youth (Chan 2010b). Those migrants generate continuous demand for low-cost housing, which is available primarily in *chengzhongcun*. The scale of *chengzhongcun* housing in Shenzhen is probably the most extreme in China, since the city was developed almost totally from scratch, and its extensive land expansion swallowed many pre-existing villages. The short history of Shenzhen's rise to prominence also means the city lacks other traditional urban social spaces that are common in Chinese cities, such as the *danwei* quarters and old city neighborhoods. However, in Shenzhen *chengzhongcun* provide a large stock and great diversity of low-cost housing, which outperform the many kinds of housing choices in Chinese cities to accommodate migrants and other low-income earners. In addition, as in other Chinese cities, the low-cost housing available in any particular form is a complex package of goods and services that extends well beyond the shelter provided by the dwelling itself.

Low-income migrants are segregated in other major cities of China, including Beijing (Zheng et al. 2009), Shanghai (Li and Wu 2008), Guangzhou (He 2013), and Nanjing (Liu and Wu 2006). Compared to Shenzhen, these cities display less extreme socio-demographic profiles (such as having large numbers of poor local laid-off workers and less skewed age structures) and housing structures (involving other types of poverty neighborhoods such as dilapidated inner-city neighborhoods and run-down work unit housing). However, the underprivileged groups are also segregated in so-called poverty neighborhoods (Zheng et al. 2009; Liu et al. 2010). These neighborhoods are pockets that accommodate low-income households and mitigate the exclusion of prime urban locations. Urban renewal programs that have become prevalent across China are in fact eradicating these more inclusive pockets in the inner city. This process is facilitating the undergoing spatial transformation characterized by rapid residential and job spatial decentralization for low-income workers and the centralization of the residential distribution of high-income households (Li 2010).

Evidence shows an alarming sign that the lowest strata of the society are being driven out of the central districts of cities and even the cities as a whole. On top of the dual social structure, the invisible hand of the market has increased segregation at a greater geographical scale and established an invisible wall around the urban core of China's big cities. To investigate the process more closely, future research should examine the changing patterns of population distribution due to the redevelopment of *chengzhongcun* and other types of low-income neighborhoods. Moreover, it would be useful to explore to what extent migrants' access to employment opportunities and urban resources are impacted by the displacement and relocation as a result of large-scale urban renewal programs.

Acknowledgement

I am grateful to Professor David Wong for valuable discussion on segregation measures, and to Professor Si-ming Li and Professor Shenjing He for their suggestions on an earlier draft. I appreciate the instructive and detailed comments given by the two reviewers and Professor Kam Wing Chan.

Disclosure statement

No potential conflict of interest was reported by the author.

Funding

This work was supported by the National Natural Science Foundation of China [grant number 41401167]; Hong Kong Baptist University [grant number FRG1/12-13/072].

Notes

1. *Chengzhongcun* are created by the land expropriation process for urban expansion, in which the farmland of peri-urban villages is requisitioned and used for new urban development, while the residential areas of the villages are retained by the indigenous villagers. *Chengzhongcun* are also referred to as villages-in-the-city (Chung 2009; Lin, de Meulder, and Wang 2011) and urban villages (Zheng et al. 2009; Wu, Zhang, and Webster 2013). However, they differ from the urban planning and urban design concept of the 'urban village,' which, in the context of Western countries, refers to a village-style urban neighborhood. Please see Chung (2010) for a comprehensive review of this topic.
2. In this paper, formal urban housing is used as a generic term for all types of housing provided by private and public sectors under urban governance. These include commodity housing, subsidized housing and dormitories.
3. 'Big Push industrialization' describes the way some countries (e.g. Soviet Union, Japan, and China) were able to generate rapid industrialization to catch up with the West in the twentieth century. It was carried out by constructing all of the elements of an advanced economy including steel mills, power plants, factories, cities, and so on.
4. Interview with officials at Shenzhen Urban Planning Bureau in July 2011.

References

Bach, Jonathan. 2010. "They Come in Peasants and Leave Citizens: Urban Villages and the Making of Shenzhen, China." *Cultural Anthropology* 25: 421–458.

Bach, Jonathan. 2011. "Shenzhen: City of Suspended Possibility." *International Journal of Urban and Regional Research* 35: 414–420.

Belsky, Richard. 2000. "The Urban Ecology of Late Imperial Beijing Reconsidered: The Transformation of Social Space in China's Late Imperial Capital City." *Journal of Urban History* 27: 54–74.

Brown, Lawrence A., and Su-Yeul Chung. 2006. "Spatial Segregation, Segregation Indices and the Geographical Perspective." *Population, Space and Place* 12: 125–143.

Chan, Kam Wing. 1994. *Cities with Invisible Walls: Reinterpreting Urbanization in Post-1949 China.* Hong Kong: Oxford University Press.

Chan, Kam Wing. 2009. "The Chinese Hukou System at 50." *Eurasian Geography and Economics* 50: 197–221.

Chan, Kam Wing. 2010a. "The Global Financial Crisis and Migrant Workers in China: 'There is No Future as a Labourer; Returning to the Village has No Meaning'." *International Journal of Urban and Regional Research* 34: 659–677.

Chan, Kam Wing. 2010b. "A China Paradox: Migrant Labor Shortage amidst Rural Labor Supply Abundance." *Eurasian Geography and Economics* 51: 513–530.

Chang, Sen-Dou. 1970. "Some Observations on the Morphology of Chinese Walled Cities." *Annals of the Association of American Geographers* 60: 63–91.

China Daily. 2009. "Shenzhen Eases Rule on Residents' Travel to Hong Kong." April 1. Accessed December 19, 2014. http://www.chinadaily.com.cn/china/2009-04/01/content_7638843.htm

Chung, Him. 2009. "The Planning of 'Villages-in-the-city' in Shenzhen, China: The Significance of the New State-led Approach." *International Planning Studies* 14: 253–273.

Chung, Him. 2010. "Building an image of Villages-in-the-city: A Clarification of China's Distinct Urban Spaces." *International Journal of Urban and Regional Research* 34: 421–437.

Chung, Him. 2014. "Rural Transformation and the Persistence of Rurality in China." *Eurasian Geography and Economics* 54: 594–610.

Duncan, Otis D., and Beverly Duncan. 1955. "A Methodological Analysis of Segregation Indexes." *American Sociological Review* 20: 210–217.

Fotheringham, A. Stewart. 1989. "Scale-independent Spatial Analysis." In *Accuracy of Spatial Databases*, edited by Micheal Goodchild and Sucharita Gopal, 221–228. London: Taylor & Francis.

Fu, Qiang, and Qiang Ren. 2010. "Educational Inequality under China's Rural–Urban Divide: The Hukou System and Return to Education." *Environment and Planning A* 42: 592–610.

Giffinger, Rudolf. 1998. "Segregation in Vienna: Impacts of Market Barriers and Rent Regulations." *Urban Studies* 35: 1791–1812.

Hao, Pu, Richard Sliuzas, and Stan Geertman. 2011. "The Development and Redevelopment of Urban Villages in Shenzhen." *Habitat International* 35: 214–224.

Hao, Pu, Stan Geertman, Pieter Hooimeijer, and Richard Sliuzas. 2012. "The Land-use Diversity in Urban Villages in Shenzhen." *Environment and Planning A* 44: 2742–2764.

Hao, Pu, Stan Geertman, Pieter Hooimeijer, and Richard Sliuzas. 2013a. "Spatial Analyses of the Urban Village Development Process in Shenzhen, China." *International Journal of Urban and Regional Research* 37: 2177–2197.

Hao, Pu, Pieter Hooimeijer, Richard Sliuzas, and Stan Geertman. 2013b. "What Drives the Spatial Development of Urban Villages in China?" *Urban Studies* 50: 3394–3411.

He, Shenjing. 2013. "Evolving Enclave Urbanism in China and Its Socio-spatial Implications: The Case of Guangzhou." *Social & Cultural Geography* 14: 243–275.

He, Shenjing, Fulong Wu, Chris Webster, and Yuting Liu. 2010. "Poverty Concentration and Determinants in China's Urban Low-income Neighbourhoods and Social Groups." *International Journal of Urban and Regional Research* 34: 328–349.

Huang, Youqin. 2004. "From Work-unit Compounds to Gated Communities: Housing Inequality and Residential Segregation in Transitional Beijing." In *Restructuring the Chinese City: Changing Society, Economy and Space*, edited by Laurence Ma and Fulong Wu, 172–198. London: Routledge.

van Kempen, Ronald, and A. şule, Özüekren. 1998. "Ethnic Segregation in Cities: New Forms and Explanations in a Dynamic World." *Urban Studies* 35: 1631–1656.

van Kempen, Ronald, and Jan van Weesep. 1998. "Ethnic Residential Patterns in Dutch Cities: Backgrounds, Shifts and Consequences." *Urban Studies* 35: 1813–1833.

Lan, Yuyun. 2005. *Dushi lide cunzhuang* [Villages in the cities]. Beijing: SDX Joint Publishing Company.

Li, Junfu. 2004. *Chengzhongcun de gaizao* [Renovation of Urban Villages]. Beijing: Science Press.

Li, Si-ming. 2010. "Evolving Residential and Employment Locations and Patterns of Commuting under Hyper Growth: The Case of Guangzhou, China." *Urban Studies* 47: 1643–1661.

Li, Si-ming. 2012. "Housing Inequalities Under Market Deepening: The Case of Guangzhou, China." *Environment and Planning A* 44: 2852–2866.

Li, Zhigang, and Fulong Wu. 2008. "Tenure-based Residential Segregation in Post-reform Chinese cities: A Case Study of Shanghai." *Transactions of the Institute of British Geographers* 33: 404–419.

Li, Si-ming, and Zheng Yi. 2007. "The Road to Homeownership Under Market Transition: Beijing, 1980–2001." *Urban Affairs Review* 42: 342–368.

Li, Si-ming, Yushu Zhu, and Limei Li. 2012. "Neighborhood Type, Gatedness, and Residential Experiences in Chinese Cities: A Study of Guangzhou." *Urban Geography* 33: 237–255.

Lin, George C. S. 2007. "Chinese Urbanism in Question: State, Society, and the Reproduction of Urban Spaces." *Urban Geography* 28: 7–29.

Lin, Yanliu, Bruno de Meulder, and Shifu Wang. 2011. "Understanding the 'Village in the City' in Guangzhou: Economic Integration and Development Issue and their Implications for the Urban Migrant." *Urban Studies* 48: 3583–3598.

Liu, Yuting, and Fulong Wu. 2006. "Urban poverty Neighbourhoods: Typology and Spatial Concentration Under China's Market Transition, a Case Study of Nanjing." *Geoforum* 37: 610–626.

Liu, Yuting, Shenjing He, Fulong Wu, and Chris Webster. 2010. "Urban Villages Under China's Rapid Urbanization: Unregulated Assets and Transitional Neighbourhoods." *Habitat International* 34: 135–144.

Lo, Chor Pang. 1994. "Economic Reforms and Socialist City Structure: A Case Study of Guangzhou, China." *Urban Geography* 15: 128–149.

Logan, John R., Yanjie Bian, and Fuqin Bian. 1999. "Housing Inequality in Urban China in the 1990s." *International Journal of Urban & Regional Research* 23: 7–25.

Ma, Laurence J. C., and Biao Xiang. 1998. "Native Place, Migration and the Emergence of Peasant Enclaves in Beijing." *The China Quarterly* 155: 546–581.

Madrazo, Brenda, and Ronald van Kempen. 2012. "Explaining Divided Cities in China." *Geoforum* 43: 158–168.

Marcuse, Peter. 1997. "The Enclave, the Citadel, and the Ghetto: What has Changed in the Post-Fordist U.S. City." *Urban Affairs Review* 33: 228–264.

Musterd, Sako. 2005. "Social and Ethnic Segregation in Europe: Levels, Causes, and Effects." *Journal of Urban Affairs* 27: 331–348.

Peach, Ceri. 1998. "South Asian and Caribbean Ethnic Minority Housing Choice in Britain." *Urban Studies* 35: 1657–1680.

People's Daily. 2004. "Shenzhen: Urban Villages Hamper Development and Progress thus Must be Redeveloped." October 26, 2004. p. A5. Accessed December 19, 2014. http://www.people.com.cn/GB/shehui/1063/2943040.html

Po, Lanchih. 2012. "Asymmetrical Integration: Public Finance Deprivation in China's Urbanized Villages." *Environment and Planning A* 44: 2834–2851.

Shenzhen Public Security Bureau. 2005. *Guanyu woshi chengzhongcun changzhu renkou he zhanzhu renkou shuju qingkuang de baogao* [Report on the data of permanant and temprory residents in urban villages in Shenzhen]. Shenzhen: Shenzhen Public Security Bureau, unpublished document No. [2005] 113.

Shenzhen Public Security Bureau. 2009. Shenzhen Household Registration Database. Shenzhen: Shenzhen Public Security Bureau, unpublished data in digital format.

Shenzhen Statistics Bureau. 2009. *Shenzhen Statistical Yearbook 2009*. Beijing: China Statistics Press.

Shenzhen Statistics Bureau. 2010. *Shenzhen Statistical Yearbook 2010*. Beijing: China Statistics Press.

Shenzhen Urban Planning Bureau. 2005. *Shenzhen shi chengzhongcun (jiucun) gaizhao zhongtiguihua gangyao (2005–2010)* [Master plan of urban village redevelopment 2005–2010]. Shenzhen: Shenzhen Urban Planning Bureau, unpublished document.

Song, Yan, Yves Zenou, and Chengri Ding. 2008. "Let's Not Throw the Baby Out with the Bath Water: The Role of Urban Villages in Housing Rural Migrants in China." *Urban Studies* 45: 313–330.

Southern Metropolis Daily. 2011. *There Is No Chengzhongcun in the Future*. Beijing: Chinese Democracy and Legal System Press.

The Guardian. 2014. "Inside Shenzhen: China's Silicon Valley." June 13. Accessed December 19, 2014. http://www.theguardian.com/cities/2014/jun/13/inside-shenzen-china-silicon-valley-tech-nirvana-pearl-river

Wang, Yaping. 2004. *Urban Poverty, Housing and Social Change in China*. London: Routledge.

Wang, Mark Y., and Jiaping Wu. 2010. "Migrant Workers in the Urban Labour Market of Shenzhen, China." *Environment and Planning A* 42: 1457–1475.

Wang, Yaping, Yanglin Wang, and Jiansheng Wu. 2009. "Urbanization and Informal Development in China: Urban Villages in Shenzhen." *International Journal of Urban and Regional Research* 33: 957–973.

Wang, Donggen, Fei Li, and Yanwei Chai. 2012. "Activity Spaces and Sociospatial Segregation in Beijing." *Urban Geography* 33: 256–277.

Wong, David W. S. 2003. "Spatial Decomposition of Segregation Indices: A Framework Toward Measuring Segregation at Multiple Levels." *Geographical Analysis* 35: 179–194.

Wong, David W. S. 2004. "Comparing Traditional and Spatial Segregation Measures: A Spatial Scale Perspective." *Urban Geography* 25: 66–82.

Wong, David W. S., H. Lasus, and R. F. Falk. 1999. "Exploring the Variability of Segregation Index D with Scale and Zonal Systems: An Analysis of Thirty US Cities." *Environment and Planning A* 31: 507–522.

Wu, Fulong. 2009. "Land Development, Inequality and Urban Villages in China." *International Journal of Urban and Regional Research* 33: 885–889.

Wu, Fulong, and Zhigang Li. 2005. "Sociospatial Differentiation: Processes and Spaces in Subdistricts of Shanghai." *Urban Geography* 26: 137–166.

Wu, Fulong, Fangzhu Zhang, and Chris Webster. 2013. "Informality and the Development and Demolition of Urban Villages in the Chinese Peri-urban Area." *Urban Studies* 50: 1919–1934.

Xinhua News. 2011. "Shenzhen Population Passes 15 Million, Non-*hukou* Population 12.8 Million, Reforms in Household Registration Next Year." December 6. Accessed December 7, 2014. http://news.xinhuanet.com/local/2011-12/06/c_122382024.htm

Yeh, Anthony G. O., Xueqiang Xu, and Huaying Hu. 1995. "The Social Space of Guangzhou City, China." *Urban Geography* 16: 595–621.

Zacharias, John, and Yuanzhou Tang. 2010. "Restructuring and Repositioning Shenzhen, China's New Mega City." *Progress in Planning* 73: 209–249.

Zhang, Li. 2005. "Migrant Enclaves and Impacts of Redevelopment Policy in Chinese Cities." In *Restructuring the Chinese City: Changing Society, Economy and Space*, edited by J. C. Laurence Ma and Fulong Wu, 218–233. New York, NY: Routledge.

Zhang, Li, Simon X. B. Zhao, and J. P. Tian. 2003. "Self-help in Housing and Chengzhongcun in China's Urbanization." *International Journal of Urban and Regional Research* 27: 912–937.

Zhang, Mingqiong, Cherrie J. Zhu, and Chris Nyland. 2012. "The Institution of Hukou-based Social Exclusion: A Unique Institution Reshaping the Characteristics of Contemporary Urban China." *International Journal of Urban and Regional Research* 38: 1437–1457.

Zhao, Pengjun, and Philippa Howden-Chapman. 2010. "Social Inequalities in Mobility: The Impact of the Hukou System On Migrants' Job Accessibility and Commuting Costs in Beijing." *International Development Planning Review* 32: 363–384.

Zheng, Siqi, Fenjie Long, Cindy C. Fan, and Yizhen Gu. 2009. "Urban Villages in China: A 2008 Survey of Migrant Settlements in Beijing." *Eurasian Geography and Economics* 50: 425–446.

Vang, Zoua M. 2010. "Housing Supply and Residential Segregation in Ireland." *Urban Studies* 47: 2983–3012.

Space to maneuver: collective strategies of indigenous villagers in the urbanizing region of northwestern China

Jing Song[a,b]

[a]Department of Sociology, Hong Kong Baptist University, Kowloon Tong, Hong Kong;
[b]Department of Sociology, Hong Kong Shue Yan University, North Point, Hong Kong

The national project of opening up Western China has led to urban sprawl in China's inland areas. However, limited attention has been paid to the localization of urbanization in Northwestern China, or the means by which villagers express their indigenous economic power by contesting spatial forms of development. This study builds on research on local collectivism to investigate strategies used by grassroots collectives to reconstruct the spatially bounded economic power of indigenous residents and create space to maneuver in the new urban landscape. Among these strategies are self-development projects, such as the construction of community housing and agricultural greenhouses. These spatial strategies are not only shaped by the distinctive policy and market environment of Northwestern China, but they vary within the region. This study investigates two inland villages in the outskirts of Yinchuan. At the time of the study, one of the villages had been absorbed into the urban area, and the other was farther away from the urban center and thus less affected by land development and urban sprawl. The findings suggest that the collective strategies implemented in villages in Northwestern China in response to land development vary with the strength of grassroots leadership, the desirability of the place, and the composition of incoming residents.

Research question

A national strategy to "open up western China" (*xibu dakaifa*) was initiated in 2000 and has since attracted much interest from researchers (Gaubatz 2008; Han and Qin 2005; Pannell 2011; Pannell and Schmidt 2006; Walcott 2003). The resulting intensification of rural land use in Western China, with an emphasis on investment in infrastructure and the expansion of economic sectors, was a new contributor to China's urban sprawl. The formation of a new urban frontier encouraged researchers to revisit the long-standing issues of social inclusion and exclusion with specific attention to urban–rural boundaries and the disparity between migrants and indigenous villagers. On the one hand, an "invisible wall" between the countryside and the city remains in place for both rural migrants and native villagers (Chan 1994). On the other, native villagers enjoy the social and economic rights anchored by China's household registration (*hukou*) system to their places of residence, whereas newcomers are further marginalized by their *hukou* status (Chan 2010a; Chan and Buckingham 2008; Chung 2013).

The disparities between urban and rural, local and migrant are complicated by regional inequalities. Although China's eastern coast has benefited from preferential policies promoting development since the post-Mao reform (Hayter and Han 1998; Veeck 1991), the nation's regional disparities have their roots in Mao's Third Front Movement in the 1950s. The nation was divided into geo-military areas based on their relation to the potential war fronts, and the Third Front region refers to Northwest and Southwest China with large-scale investment as military industrial centers (Naughton 1988). In a pattern of managed development, state-initiated programs have resulted in distinctive combinations of residential inclusion and social exclusion, such as state farms (Hansen 2004) and military agricultural corps (*bingtuan*) (Matthew and Cliff 2009). This study revisits these issues in the latest context of rapid urbanization and accelerating population migration, given the government's efforts to reduce regional disparities, and ensure that inland cities reach the standards set by their eastern counterparts (Gaubatz 2008) and secure a "harmonious society" (Yep 2013). In China's new policy and market environment, with its emphasis on modernization and beautification, the recent intensification of land use and the influx of migrants have posed new challenges for inland areas that lag behind in terms of providing specialized services, high-quality infrastructure and access to international markets (Walcott 2003). In Northwestern China in particular, land and housing markets are less mature and have not attracted growth-inducing investment comparable to that in other areas. This provides a natural laboratory in which to examine the process of rapid urbanization in a sparsely populated and less-developed area, and the means by which locals negotiate their living space and attain spatially bounded economic power.

The phrase "spatially bounded economic power" highlights the influence of space on power relations and the pursuit of social justice (Chung 2013), given that administrative boundary remains powerful in "bounding" economic activities (Chan 2010b). Hsing (2010) affirmed the significance of jurisdictional boundaries to the organization of local economy with reference to Liu's (2004) theory of "jurisdictional economy," and Smith (2014) identified "territory" as a form of currency common to a range of actors developing strategies and negotiating in response to village transformation. Based on the grassroots conceptualization of power, indigenous villagers and collectives claim specific social and economic rights associated with their "local citizenship" (Smart and Smart 2001). The resulting local spatial freedom provides indigenous villagers with a distinct "space to maneuver" in response to rapid urban sprawl, helping them resist top-down urban expansion, negotiate better deals, and redirect existing plans.

Such spatial freedom needs to be contextualized in China's unique institutional structure of five-level local governance: the province, prefecture, county, township, and village.[1] In the spatial hierarchy of local governance, each level is theoretically subservient to its superior government, but local officials also have relational discretion such as regarding economic policy. Meanwhile, urban governments exercise administrative authority over rural areas, and this unequal power relationship played an important role in negotiating land expropriation between urban (municipal and district) governments and rural (township and village) governments (Hsing 2010). This paper uses the term of "local governments" to refer to governments at the municipality and district levels that initiated land development projects, and they established a "development office" together to supervise the development plans directly. With urban expansion, several districts were expanded to include new subdistricts to replace the dissolving rural townships. But at the village level, some village committees remained functional as the direct collective owner of the to-be-developed land in the transitional period. In this paper, "grassroots leadership" refers to the leadership of rural collectives at the levels of villages and villagers' groups, who were the most important collective actors, who responded to the development projects.

Previous studies of urban sprawl and village transformation have shown that discretion is permitted in the implementation of development plans to accommodate local circumstances, as reflected in Guangzhou's "one village, one policy" initiative (Wu, Zhang, and Webster 2012). With increasing localization of development strategies (Ma 2005), land development in inland areas is expected to show a great diversity, despite the presence of a common set of overarching institutions responsible for development planning and implementation. The aim of this study is to determine how grassroots strategies and reactions to large-scale development projects express indigenous economic power in Northwestern China. The localization of development policies is examined not only in the distinctive context of Northwestern China, but with specific reference to two villages in the same outlying area at different distances from an expanding urban center. This study investigates grassroots strategies in response to the "Greater Yinchuan" development project (focusing on the capital of the Ningxia Hui Autonomous Region), based on analysis of government documents and in-depth interviews in two villages. At the time of the study, one of the villages had been converted into a district of Yinchuan; the other was farther away from the urban area but also influenced by urban sprawl.

Urban sprawl, land development, and local responses

Macro-level strategies relating to urban sprawl and land development in China evolved from the hurried creation of development zones in the 1990s (Deng and Huang 2004; Liang and Zhou 1993; Peng 2001) to the construction of "new cities" in the 2000s. The latter projects emphasized property management, rather than industrial investment (Hsing 2010). China's leading municipalities have hosted major events, such as the Olympics (Beijing in 2008) and the World Expo (Shanghai in 2010), to enhance the nation's modern, orderly, and outwardly oriented urban image (Gaubatz 2005, 2008; Smith and Himmelfarb 2007; Wu 2000a, 2000b). The implementation of a modernization campaign extending from the coast to inland areas as part of a new national strategy to develop Western China has caused a massive spatial reorganization in the form of urban sprawl into the countryside. Officials have joined the rush to approve the construction of factories, office towers, and apartment buildings on agricultural land to achieve their ambition during their term in office (Ho 2001; Zweig 2000). Property rights in cities are becoming increasingly commercialized, and the privatization of housing is accompanied by rising housing inequalities and market-driven residential segregation (Xu 2008). Research on China's reform has revealed a decentralization of command and increasing emphasis on the market mechanism (Lin 2010), whereas governments still retain strong control over development programs to a large extent.

However, the bottom-up spatial strategies that help to shape land development and place promotion are disregarded in frameworks that emphasize the strength of the state or the prosperity of the market. For their indigenous residents, village communities are settings for life and work over generations, rather than sources of revenue. However, China's aggressive land-development campaigns have frequently caused residential displacement, as low-income residents have left the most desirable places, making room for economic growth (He 2007; He and Wu 2007). In addition, such campaigns have caused grassroots conflicts, as the compensation received by locals accounts for only a small share of the profits from land sales (Deng and Huang 2004; Ren 2003), and locals wish to participate directly in the development process (Yep 2013).

In the role of individual owner developers, some villagers have initiated dense residential construction on their land, thereby generating rental incomes (Chung 2013; Liu et al. 2010; Zhang, Zhao, and Tian 2003). In addition, collective endeavors have become essential in carving out an autonomous economic sphere beyond the reach of formal development control. In many cases, the organizers of macro-level development projects do not interfere with the "minor property rights" (*xiaochanquan*) housing constructed by township governments and village committees (Paik and Lee 2012) or with the small-scale fringe development undertaken by villagers' shareholding cooperatives (Hsing 2010). Governments have also adopted a mix of cooperation and coercion to smooth the development process by offering "reserved land" to villages or allowing locals to return after temporarily vacating their homes (Hsing 2010; Wu, Zhang, and Webster 2012).

Parallels can be drawn in China today between the state's tacit approval or tolerance of grassroots initiatives – by townships, villages, or villagers' cooperatives to fight for their economic interests – and local economic experiments in the early reform period, such as the reinstatement of the family-farming system and the establishment of township and village enterprises. Many researchers have recognized the importance to rural industrialization of local collective leadership, whether in the form of "local state corporatism" (Oi 1992) or "local market socialism" (Lin 1995). Due to the decentralization of the socialist state, the rearrangement of fiscal relations, and the localization of property rights (Walder 1992), local governments can help enterprises to obtain materials, licenses, and tax savings and maintain control of resource allocation and investment decisions (Oi 1989). Much less attention has been paid by researchers to recent developments in local collectivism as a result of urban sprawl. In response to the encroachment of central or municipal governments, the local corporate structure may become increasingly unified, emerging as a coalition of interest groups with local imperatives. In Southern China, collective resistance to development has been facilitated by strong lineage organizations and village corporatism (Hsing 2010). An attempt is made in this paper to examine the amount of space left for local collectivism in Northwestern China and to identify ways to promote indigenous economic interests.

Previous studies have documented various expressions of indigenous economic power during urbanization and village transformation. Smith (2014) compared three transforming villages and identified different transformation initiatives created by village – development corporations (Guangzhou, Southern China), state-directed networks of investment at the township and municipality levels (Suzhou, Eastern China), and village leadership, based on local networks of kinship and entrepreneurship (Chongqing, Southwestern China). Locals have used their territory and networks to contest changing development scales and insulate their communities from socio-spatial change or to reinterpret development plans and negotiate the "best" path for transformation. In contrast, migrants have attempted to create social space in the face of exclusion. Po (2008) compared three villages in Guangdong (Southern China), Jiangsu (Southeastern China), and Beijing (Northern China) and found that various shareholding reforms implemented by rural collectives have allowed landless farmers to benefit from urbanization. Leaf (2002) compared a village integrated into the expanding urban economy of Quanzhou (Southeastern China) with a parallel case of peri-urban change in Vietnam, and discussed the "sharing" of territory between urbanizing villagers, suburbanizing urban professionals, and temporary migrants from outside the given province. Chung (2013, 2014) focused on "villages-in-the-city" in Guangzhou and found that shareholding companies develop various businesses based on their collective assets and land resources and allocate social goods to native villagers in a closed system. In Southwestern China, indigenous villagers have not only embraced "rightful-resistance" strategies, but also have sought to take advantage of the rural-to-urban transition by building housing extensions in the hope of securing larger compensation packages (Lora-Wainwright 2012).

Due to the great variation in village planning in coastal and inland areas, it is misleading to represent the whole of Western China as equally subject to the "Xibu Da Kaifa" ("Open Up Western China") imperative. The focus of this study is the relatively underresearched region of Northwestern China, a large and sparsely populated area. It makes up around 30% of the nation's land and 4% of the population, with a high concentration of ethnic minorities. Despite the presence of the country's major heavy industry projects, northwestern areas remain less "opened-up" to market penetration in many aspects. As such, the spatially bounded economic power is negotiated in a regionally specific context in two key respects.

First, the local governments of China's Ethnic Autonomous Regions enjoyed a greater autonomy with regard to development plans and compensation deals given by central authorities. The macro-level policies have been directed towards multiple commitments, including economic development, environment protection, and social stability. The central government initiated projects to conserve natural forests or convert cropland on steep slopes to forest and grassland for ecosystem services, for example, and such government interventions were often coupled with huge payments. Meanwhile, the central government also encouraged local governments to accelerate the development of the lagging infrastructure and economy, in contrast to the "cool-down" measures in the coastal areas.[2] In addition to receiving subsidies, local governments were encouraged to generate their own revenues to add to the investment. In general, the typical land conversion pattern in Northwestern China was from grassland to agricultural land and then to urban land, but in this process, multiple goals needed to be balanced, such as to maintain "social harmony" in areas with a concentration of ethnic minorities, and this required local knowledge and grassroots cooperation. Therefore, local governments are given considerable leeway to manage their own developmental plans and, at the same time, the responsibility to avoid conflicts that occurred in early-developed areas due to land development.

Second, in these sparsely populated areas, many land resources have not been commercially attractive enough for developers. Other areas of inland China (such as Southwest and Central China) were more densely populated and witnessed high levels of land commercialization and market penetration, such as Chongqing, which has experienced high-density residential development in its peripheral areas (Smith 2014), Wuhan, with its expanding commercial housing market (Wang and Fan 2012), and Kunming, with its emerging "middle-propertied" stratum (Zhang 2010). In Northwestern China, dryland farming has yielded limited productivity due to the constraints of surrounding mountains and scarce water resources, and commercial activities were restrained by the limited scales of population the area could support.[3] Lacking local-based industrialization dynamics, this area typically accommodated transplanted industries and government projects. To facilitate urban sprawl, favorable and inclusive measures are needed to attract investors and newcomers. Such unique characteristics affect how developmental projects are received and negotiated at the grassroots level.

With particular reference to Northwestern China, this study investigates the differences between villages encroached upon by cities and villages that are relatively untouched by urban development. In addition to villages subsumed into expanding cities (Chung 2013, 2014; Leaf 2002; Smith 2014), cases of villages farther from cities have been documented in the literature. These villages are subject to different development plans, such as the "New Rural Reconstruction" movement observed by Day (2013) and Hale (2013) in the northern, central, Midwestern, and southwestern provinces of China, and the "Building a New Socialist Countryside" movement illustrated by Looney (2012) in Ganzhou, Central China. The former movement is a grassroots effort stimulated by intellectuals with loose ties to the regime, and the latter movement is a government top-down program, and as suggested by the cases in this study, is nevertheless continuously modified locally. Various projects of construction, demolition, and planning included building designated "new villages" (Ahlers and Schubert 2009; Schubert and Ahlers 2012) and reconstructing a "tourist" landscape to reap indigenous economic benefits (Chio 2014 in Guangxi and Guizhou). Because these villages are farther away from cities, they experience different combinations of external and internal forces, and they face different economic links with urban markets and migration flows. Therefore, the efforts made by native villagers to negotiate their sense of ownership and autonomy in relation to outsiders must be contextualized with reference to their geographic locations in the urbanization and modernization campaign.

In sum, the diverse patterns of urbanization and village planning necessitate more nuanced categories for analysis of the localization of development policies than those provided by the conventional top-down/bottom-up framework (Smith 2014). With attention to the distinctive policy and market environment of Northwestern China, this study investigates the forms of spatially bounded economic power available to indigenous residents. The study addresses local collective strategies for creating space to maneuver in response to urban sprawl, and shows how grassroots strategies vary according to villages' geographical characteristics, particularly their distance from expanding urban centers.

Data and method

The two villages under study are located in the Ningxia Hui Autonomous Region. Both villages are near the capital of Ningxia, Yinchuan. Yinchuan occupies a historically strategic location. Today, the municipality is assuming increasing importance in the national campaign to open up West China. The administrative area of Yinchuan has a population of 1.99 million, distributed among three urban districts, two counties, and one satellite city. In 2010, Yinchuan was officially designated as the host place for the China–Arab States Cooperation Forum, enabling the municipality to become a center for business between China and the Arab States by hosting events, such as the China–Arab States Expo.

This study draws on in-depth interviews conducted between 2002 and 2014 with villagers and local officials from two villages: Ning Cun and Pan Cun (pseudonyms). The research team[4] used a snowball-sampling procedure to recruit respondents, and the interviewers selectively followed up on the solicited referrals to avoid sampling from a narrow social circle and to obtain even coverage in terms of age and socioeconomic status.[5] At the time of the study, each village was home to approximately 4000 people and 10 "villagers' small groups" (cunminxiaozu).[6] Prior to the development of their land, both villages had been predominantly engaged in dryland farming, with 3000 mu (1 mu = 666.7 m^2) of farmland in Ning Cun and 4700 mu in Pan Cun. Ning Cun has hosted a large number of sideline businesses that directly served the urban service sector, due to its proximity to Yinchuan. In contrast, Pan Cun has experienced considerable labor outmigration to distant cities, as it is near three important local highways connecting urban centers with the countryside.

Ning Cun underwent massive land development between 1992 and 2008 due to urban sprawl, and was transformed into several neighborhoods of one of Yinchuan's expanding city districts. Most of the village's land was taken away in increments to construct commodity housing, office buildings, factories, and trade centers. In return, the developers provided monetary compensation distributed at four levels (the township, the village, villagers' groups, and individual households) and offered relocation housing units in the newly developed residential areas. In the biggest villagers' group, a community housing neighborhood was built as an alternative residential option for native villagers.

Pan Cun belongs to a rural county 12 kilometers from Yinchuan. It was affected by the Greater Yinchuan project, but to a lesser extent than Ning Cun. The county government aimed to expand real estate development and the service sectors to meet the demand created by migration to the expanding municipality. Some of the land in Pan Cun has been developed for the construction and expansion of roads and other infrastructure, and the locals involved in our study expressed some anxiety about their future compensation for the consequences of urban sprawl and the influx of newcomers. Several villagers' groups mobilized villagers to invest in agricultural greenhouses and restore the locals' ownership of land and property. These grassroots spatial dynamics contributed significantly to the localization of formal large-scale development projects.

Ning Cun: modern version of self-help housing

As part of the rapid urbanization of Yinchuan's surroundings, Ning Cun was incorporated piece by piece into the new urban area. Prior to its expropriation, land in Ning Cun was collectively owned and the ownership was shared by the township, the village, and villagers' groups (called "teams" by the villagers). These teams played an important role in land management at the micro level. As more and more teams took on the responsibility for land development, developers were forced to negotiate new compensation packages. The land price offered by the developers increased from 80,000 yuan per mu in 1998 to more than 100,000 yuan per mu in 2003, and the villagers' share of the lump compensation increased from 10,000 yuan to 30,000 yuan. The land prices offered to the teams – even those whose land was developed at roughly the same time – varied due to differences in the strategic positioning of the land. These inconsistent compensation principles led Ning Cun's residents to suspect that the local cadres were pocketing some of the compensation and working against the community's interests. As a result, many villagers felt a need to conduct self-development activities, as their land or housing properties could be taken away or demolished at any time.

In response to urbanization, villagers might intensify their housing development not only to generate rental income, but also to secure greater compensation after demolition, as shown by Anna Lora-Wainwright (2012) with reference to Southwestern China. However, self-help homes have often been described in official rhetoric as poorly constructed, unregulated, and deteriorating living environments. Villagers have found it difficult to resist the official crackdown on self-help housing without collectively "upgrading" their self-developed buildings. To ensure tolerance within the government's regulatory environment, such self-help development requires more investment than that provided by individual savings, which is made possible by the collective compensation for expropriated land.

The seventh villagers' group in Ning Cun (called "Team Seven" by the villagers) was able to mobilize its members' collective compensation for this purpose around 2001. Most of the other groups had distributed this compensation to the villagers or used it to meet other expenses. The villagers in Team Seven, as in other groups, suspected the village cadres of conspiring to appropriate their collective compensation, and several opinion leaders brought a petition letter to township leaders signed by the majority of the group members. The letter stated that the township had expropriated a piece of land without following legal procedure; however, rather than requesting that the land be returned, the group asked to elect a new leader, who would use the collective budget transparently. The township government gave in. In 2001, one of the opinion leaders won such an election and proposed using the group's collective compensation to build a community housing neighborhood.

As the group's land was to be completely expropriated by local governments at the municipality and district levels, the community housing neighborhood had to be built on land relinquished to the village by local governments. The new group leaders had to apply for a piece of land once owned by a rural collective but subsequently transferred to urban governments. To gain approval from various government authorities, the new group leadership designed their development plan to "correct construction disorder and beautify the city image." Fitting well into the official narratives of investment-oriented urban planning, the plan to build a community housing neighborhood was approved by local governments at the municipality and district levels. The housing neighborhood, namely Spring Garden, was constructed in 2003. Every household in the villagers' group was eligible for a single-family house in Spring Garden, but was required to pay a deposit of 30,000 yuan to ensure that the construction could be completed on time. The remainder of the price of the house, payable on receipt of a house key, was proportionate to the amount of land remaining to the family.[7] By 2008, when most of the land-development projects had been completed, Spring Garden had become a reserve of peasants' dream houses encircled by the expanding urban area, an "upgraded" version of the illegal self-help housing characterized by single detached homes and front yards.

This spatial outcome reflected the locals' effort to retain their indigenous economic power by exchanging their farmland usage rights for family homes. Group members could choose to live in officially sanctioned relocation housing or self-developed housing, and obtained extra revenue from rental businesses and housing sales. Although local conjugal families were entitled to buy houses at minimum prices, some villagers lost this privilege by giving up their local *hukou* before Spring Garden was constructed. One of the respondents regretted his decision so much that he chose to buy an urban *hukou*, and left his land behind. As shown in previous research, villagers with greater awareness of the potential value of their land to investors are less likely to relinquish their rural *hukou* (*nongzhuanfei*) (Lora-Wainwright 2012). In some cases, collective landownership entitled locals to claim benefits from investors. Native residents, who had remained within such villagers' groups during the development project, felt fortunate that they had not missed the opportunity to negotiate a good housing deal.

Nevertheless, with the influx of newcomers due to land development and urban expansion, many outsiders bought their way into Spring Garden. Some villagers chose to sell their houses to fund their urban lives, and the huge profit gap between market prices and internal prices encouraged others to convert their housing advantage into cash – despite the possibility of this gap's increasing further in the future. Even without receiving a landholding discount on their initial purchase, locals could gain a net profit of more than 100,000 yuan by selling a house in Spring Garden, which cost 20 times the local average annual income per capita. However, some better-off families chose to buy second houses without the insiders' discount, costing approximately 100,000 yuan each. The collective was permitted to sell any houses not claimed by locals to outsiders at the market price.

By formalizing their indigenous *de facto* rights to properties, the villagers' group managed to secure their own mechanism of resource allocation. However, a subsequent campaign to modernize and beautify the locality threatened to diminish this indigenous economic power, as newcomers flocked to the newly commercially attractive Spring Garden. By 2014, around half of the houses in Spring Garden had been purchased by non-local emerging middle-class homeowners, and the indigenous community was fragmented, as many locals had converted their housing into cash and moved elsewhere.

Pan Cun: capital-intensive greenhouse project

Although Pan Cun was farther away from the outer reach of urban expansion, the village was still affected by the resulting rise in land and housing prices. Like those in the many areas in which land assets were commercialized, the village leaders expected eventually to "sell" their land and the villagers to become "urban residents" (Zhang 2000). In 2007, land was expropriated on a small scale to enable infrastructure construction, and villagers received compensation of approximately 15,000 yuan per mu, much less than the compensation received by villagers in Ning Cun. The remaining collectively owned rural land was still expected to be used for farming and residential purposes only, and the transfer of land use had to be approved by higher level authorities. Without legal access to capitalize on either land or housing, the native villagers began to develop illegal self-help housing as well, but lacked the financial capability and official license to transform the construction of individual homes into a collective project. Furthermore, as Pan Cun was merely a stopover for many rural migrants traveling to Yinchuan from poorer areas, the rental businesses were not as lucrative as in Ning Cun.

Many locals joined the migration to the cities for work. Although it was easy for young men to find construction work in the expanding urban areas and for young women to find jobs in the service sector, migrants had little opportunity to build lifelong careers; except a few with sufficient education to obtain white-collar jobs, many of the respondents spent only 3–4 months each year working on the cities' construction sites. Although the construction projects provided migrant workers with decent wages (about 100 yuan per day), the workers were easily replaceable and the jobs often physically demanding. In the words of an elderly respondent, "If a person of my age were to do migrant work, he would soon die in the workplace." Nevertheless, it was necessary for the locals to continue migrating between Pan Cun and Yinchuan to ensure a sustainable income. The accumulation of capital allowed some locals to start up small businesses, and a few became economically well established as construction contractors. In contrast, agricultural work had a profit level of approximately 500 yuan per mu, which was only just enough to ensure self-sufficiency. As a result, people did not take their land entitlement seriously. Young people tended to leave their land in the hands of the elderly. Some land was left idle and was taken over by relatives, neighbors, or other villagers with the permission of the villagers' groups.

Things changed when several villagers' groups ("teams") began collectively to organize the construction of agricultural greenhouses. The project was initiated by the township government in 2001, but was not effectively implemented due to a lack of financial investment. In the following years, several teams mobilized sufficient resources to take over the leadership of the greenhouse project. The teams received some subsidy from the township government; they were responsible for obtaining the remainder cost from individual households. Team One was the first to solicit input and technical support for the greenhouse project. Teams Three and Four joined the project in around 2005 and established greenhouses on a larger scale in the following years. Farmland was converted into small or large greenhouses ranging from 10 to 40 m in length. The land owned by individual households (in terms of the use right) was either partially or completely converted into greenhouses, depending on the collective zoning plans. The agricultural output, which was mainly vegetables, rather than the regular dryland crops of wheat and corn, was collected at the community level and transported to the urban market. The neighboring counties also collaborated to establish a unified center for trade in local vegetables and other agricultural products, with a total investment of 940 million yuan in a development zone of 655 mu. The organizers stated that their priority was to create employment opportunities for local peasants.

The development of greenhouses generated comparable but more stable returns compared with migrant work. Many migrant workers returned to Ning Cun to claim their right to farmland before the project had begun. In restructuring the value of local land, the greenhouse project was also an attempt to retain local ownership of land in the face of urban sprawl. Some respondents intentionally grew certain economically valuable crops or fruit trees that "would be compensated at a high price" in case of land expropriation. By 2009, the greenhouses constructed by Teams Three and Four had expanded from 1000 mu to 1500 mu, covering more than half of their farmland. In some cases, the annual income derived from the greenhouses was as high as 10,000 yuan per mu, although market fluctuation continued to introduce risks. More teams joined Pan Cun's greenhouse project in 2009, namely Teams Two, Five, and Six.

The organizers of the greenhouse project offered favorable deals to insiders. The teams were successful in obtaining financial aid from the township government to support the construction of greenhouses by individual households. Some peasants received 200-yuan subsidies for building small greenhouses costing approximately 1000 yuan each, and others received 4000-yuan subsidies for building large greenhouses at 20,000 yuan each. Villagers were permitted to receive subsidies for as many greenhouses as they could afford to build on their farms that were included in the "greenhouse zone." They were moderately compensated for returning the remainder of their farmland to the collective. Locals who could afford to build more greenhouses were permitted to request more land from the collective. Greenhouses were constructed on any collectively owned land that remained and rented at a higher rate to outsiders. If the local residents encountered financial difficulties, they were permitted to sell their greenhouse usage rights to outsiders at a market price of approximately 50,000 yuan per greenhouse. The land beneath these rented greenhouses was still owned by the collective, who received rent for the land from the outsiders. Alternatively, locals were permitted to rent their greenhouses to outsiders for 5000 yuan per year, which was still much more lucrative than renting an equivalent amount of regular farming land (700 yuan per mu each year). Therefore, the greenhouse project was also a means of reconstructing indigenous economic power and "individualizing" the benefits derived from collective land entitlement. In distributing such spatially bounded economic resources, similar to what happened in Ning Cun, the initial distribution was limited to insiders in an egalitarian manner, but the collective allowed better-off families to claim more resources from others and gave outsiders the lowest priority.

Nevertheless, the greenhouse project attracted newcomers in search of employment opportunities. Many were from rural areas in the neighboring counties and provinces, and rented greenhouses either from the collective or from individual households. They considered it more prudent to settle down in Pan Cun than to search for long-term work in cities. The incomers were ideal tenants for the indigenous landlords; some even arrived with their families and bought new homes in the villages.[8] Compared with Ning Cun, Pan Cun provided a relatively loose environment for rural migrant workers to work in greenhouses, rent or buy rural housing, and send their children to local schools. This flexibility was in part due to real estate developers' lack of interest in the village's land. The cost of building one's own conjugal home in Pan Cun was only 40,000 yuan, whereas a commercial housing unit with equivalent floor space cost 150,000 yuan in the center of the local county, and cost more within the urban area of the municipality into which Ning Cun had been incorporated.

The prices of land and housing in Pan Cun were expected to rise in the long term as a result of the land development fever, dubbed as "land-enclosure movement"[9] by some (Deng and Huang 2004). The grassroots cadres did not have the power to approve land- or housing-development plans, and had limited leverage when negotiating with higher authorities. However, the conversion of farmland to greenhouses gave local villagers the opportunity to claim a larger share of development revenues. The greenhouse project was more capital intensive than the regular farming sector and benefited from closer connections with urban markets. The collectively mobilized greenhouse project was also a means of retaining ownership of land to reinforce villagers' economic entitlement; the locals expected greater compensation if their land was subject to future development plans.

Grassroots leadership: undermined or empowered?

Grassroots strategies with the support of the local collective leadership were implemented in both villages in response to urban sprawl. Approaches other than the top-down/bottom-up framework should be used to examine these grassroots initiatives, which provided innovative means of accessing resources and mobilizing networks that cannot simply be categorized as unregulated and spontaneous (Smith 2014). Although local individual developers were excluded from the top-down zoning practices, their input was solicited by the intermediate agencies of township governments and village cadres. In both villages, efforts were made at a grassroots level to attain a degree of ownership and autonomy recognized by official agencies or incorporated into the official place-making process. The villages' collective ownership of economic resources provided indigenous residents with more favorable deals than those available to outsiders in the fledgling open market. In this process, the spatially bounded economic power and the "space to maneuver" was largely shaped by the strength and structure of the villages' grassroots leadership.

The organizational capacities of the grassroots leadership were subject to the changing structure of resource control. Under the ambitious development plan of the Greater Yinchuan project,[10] the development office tended to monopolize the development initiatives and to put development-related resource coordination under its direct supervision. Ning Cun was among the first ones to be urbanized, and its economic resource management was largely taken over along with the transfer of collective land entitlements.[11] Grassroots alliances were only mobilized at the level of villagers' groups, whose cadres and opinion leaders had a direct interest in promoting indigenous economic power. However, widespread urbanization also created an opportunity for grassroots empowerment because various parties were required to work out land transfer deals in a short time. The municipal and district governments were eager to complete land expropriation efficiently and "without further trouble," according to a local official (Interview with author, 26 June 2014).

Furthermore, as social harmony and political stability were key criteria for evaluating the performance of local officials, the local governments used compromises to "compensate" for their "mistakes" for the sake of social stability. During the process when local governments expropriated land from rural collectives and leased land to developers, they sometimes failed to balance the pace of land requisition with that of land development. When the expropriated land was not developed in time – within "two years," in official terms – local governments faced official scrutiny for potential misconduct.[12] Similar "illegal" and "amoral" government conduct included expropriating more land than planned in the name of public interest to generate income from rents and taxes. Such governmental misconduct gave grassroots collectives – namely the owners of rural land – greater scope and leverage with which to bargain with the authorities. In Ning Cun, villagers' groups also intervened in the distribution of employment opportunities generated by land development. When a piece of land owned by a particular group was developed, the members of that group benefited from employment opportunities at the local construction sites, and were paid for transporting the construction materials to pave the land for subsequent development. However, such compromises only existed in the interim period, during which governments and developers sought desperately to secure the cooperation of grassroots collectives to facilitate macro-level zoning. For example, the Spring Garden project was initiated by Ning Cun's biggest villagers' group, which owned large tracts of land and thus gained considerable bargaining power in the transitional period of urbanization.

In Pan Cun, the grassroots leadership retained relatively strong control of local economic resources and the structure of rural governance remained largely intact. The rural land remained under the collective ownership, and grassroots leaders enjoyed greater freedom to shape the local landscape. However, as local governments were reluctant to monopolize development initiatives in the locality, Pan Cun's grassroots collectives did not have the same bargaining power as their counterparts' in Ning Cun. As in other villages, the residents of Pan Cun were only permitted to use rural land for farming and residential purposes, and failed to obtain official permission for collective housing development. To make the most out of the dryland agriculture, the greenhouse project was first initiated at the township level, and was mobilized by villagers' groups as the best means of generating capital-intensive investment, which may strengthen their future bargaining power. It seems that the respondents were waiting for the moment of large-scale expropriation, which would create a good chance to carve out their own sources of indigenous economic power in the expanding urban economy.

In sum, the grassroots leadership in Pan Cun remained largely unchallenged, but lacked the leverage to participate in or alter the implementation of top-down zoning plans. In Ning Cun, the local collective leadership was undermined by the rapid and widespread process of urbanization, but the cooperation of the village's grassroots collectives provided them with space to maneuver within the officially monopolized regulatory environment. During this transitional period, grassroots leadership had a window opportunity to bargain for a greater share in development revenues because the municipal and district governments were very concerned about building harmonious social and ethnic relationships in these areas. But over time, such gains might be undermined after land development was finished,[13] and the spatially bounded economic power available to the native villagers was also penetrated by external forces as a result of market dynamics.

Market dynamics: attractiveness and competitiveness

With regard to market dynamics, northwestern areas have not attracted growth-inducing investments comparable to those in other areas. Built on the historical legacies of dry-land farming, villagers had been locked in agriculture and circular migration, and the two villages lacked local-based entrepreneurial motivations to industrialize and urbanize the locality. As such, favorable measures should be implemented to add to the locational attractiveness, to invite more developers and outsiders to come, and to reverse the migration flow. The sudden exposure to the government-initiated urbanization process aroused both anxiety and hope among villagers worrying about losing everything to predatory governments and exploitative developers, on the one hand, and envying those, who were generously compensated for land expropriation, on the other. This means that villagers did not want to simply resist and turn down the development opportunity, but were willing to seek cooperative development plans that could take care of their economic interests. Furthermore, villagers were not sure about the "actual" prices of their properties, and many regretted selling their land or housing "too early." Facing the uncertain future, local people began to explore ways to retain some control, if not all, over local economic resources.

Ning Cun's collective housing project reflected how villagers and rural collectives were in a process of learning how to cope with the sudden urbanization project. Compared with Pan Cun, Ning Cun has been more commercially penetrated due to its proximity to the urban center, and many villagers have been exposed to the market dynamics of small businesses. As the development plans unfolded, places near the urban center or in otherwise strategic positions, such as Ning Cun, were increasingly regarded as attractive. Many of the houses in Spring Garden were sold at a price comparable to a suburban villa house and generated considerable revenues for villagers. However, some villagers commented that they would like to hold on to their properties to generate more sustainable income via rental business, but they were forced to sell their houses to solve their urgent financial need and to meet the demand of urban life. Moreover, villagers were not sure about the sustainability of such a property boom. This lack of confidence in the locational commercial desirability made it logical to sell their properties when the place was still somewhat attractive.

Pan Cun was less commercially desirable than Ning Cun due to its greater physical distance from the urban market, and thus, the village's grassroots collectives had more time to prepare for land expropriation. Rather than waiting passively, grassroots collectives in areas far away from Yinchuan were able to initiate their own development projects and thereby spatially reorganize local property rights and economic resources. Many locals, who had previously migrated from the village to work, returned to claim their spatially bounded economic power and secure their right to potential compensation in future waves of urbanization.

However, due to Pan Cun's less-desirable location, the locals' gatekeeping activities were insufficient to reap economic benefits and defend indigenous economic power during urbanization. It required more investment to make the village attractive, compared with villages already encroached upon by the municipality. Pan Cun's villagers' groups converted dryland farms into technology-based vegetable greenhouses that became part of Yinchuan's growing urban network of supply and demand. Compared with Ning Cun, in which the profits derived from self-developed housing were more or less guaranteed by the village's proximity to the urban center, people in Pan Cun were less confident about the rewards of their land investment. As a result, some villagers opted out of the collective greenhouse project, especially young people who found good urban jobs. Their parents often took over the responsibility of "watching" their land to prevent its being taken back by the collective. Therefore, although some land and greenhouses were rented to outsiders, the rights to the land were retained by indigenous residents or collectives, who expected the value of land to rise as a result of urban sprawl. However, their self-developed housing could not be rented at rates comparable to those properties' in Ning Cun, and their hope was to attract more potential tenants and to approve the value of their properties when urban expansion continued. In short, the grassroots development projects in Pan Cun to reconstruct spatially bounded economic power were also designed to make the village attractive, and were thus more inclusive and tolerant of migrants than in Ning Cun. Even in Ning Cun, to consolidate indigenous economic power, the village's attractiveness had to be maintained through continuous investment, such as the payment of fees for managing utility maintenance and service delivery.[14]

In sum, the spatially bounded economic power of locals in Pan Cun and Ning Cun was determined by the attractiveness of their respective localities. Grassroots collectives also implemented inclusive measures to make their villages more desirable, such as leasing greenhouses and renting or selling housing properties to outsiders. About a third of Pan Cun's residents were newcomers, most of who had recently moved from other rural areas as agricultural laborers. In Ning Cun, the newcomers came from both urban and rural areas, and they were economically more established and looked for affordable housing in a suburban area. These newcomers posed a threat to the locals' indigenous economic power because they bought houses as new homeowners independent of the villagers' group.[15] In contrast, newcomers to Pan Cun usually gained access to greenhouses or self-built residences on a lease basis. They did not compete with locals directly in the market, but continued to generate rental income for them.

Due to their different levels of attractiveness and compositions of newcomers, the two villages created space to maneuver in different ways. In Ning Cun, the local residents received greater rewards for their collective investment in housing assets due to the attractiveness of the village, but it became more difficult for villagers to act as a collective due to their weakening leadership and greater market penetration from the outside. In Pan Cun, the structure of rural administration remained intact, but villagers' groups were still seeking better ways to negotiate better deals with governments and developers, and they implemented more inclusive measures to allow immigrants to participate in the spatial reorganization of local economic resources.

Concluding discussion

The strategies implemented by grassroots collectives in Ning Cun and Pan Cun offer new insights into urbanization outcomes, especially in Northwestern China, whose distinctive socioeconomic characteristics offer a natural "laboratory" for research on urbanization. The unprecedented pace of land expropriation has dispossessed many villagers, but also created space for grassroots self-development projects to restructure and maximize indigenous economic power. The grassroots initiatives documented in this study may be part of a wider experimentation with rural land rights supported by local collectivism. Although the state has yet to recognize or regulate these activities, local collectivism may suggest new directions for China's reform. In the absence of formal legal directives to empower indigenous residents, people began taking action via self-help development organized by local collectives.

The observed spatial strategies implemented at the grassroots level in response to land development were shaped by the distinctive policy and market setting of Northwestern China. First, the less-than-stringent developmental controls and ambiguous development plans made it possible to reinterpret development plans at the grassroots level and to create space to maneuver by rural collectives. In a top-down zoning model, the municipality government and private developers reap the greatest rewards. However, due to the recent official emphasis on social harmony and the competing commitments of local governments, some leniency was considered necessary to facilitate and smooth land expropriation, which empowered grassroots collectives in negotiation and allowed for more cooperative measures to be adopted in development projects.

Second, the region's land and housing markets had not attracted comparable real estate and commercial interest as in other early-developed areas. Villagers were less confident about the property boom, and they struggled between the desires to sell their properties and to keep a close control over their economic resources. The future uncertainty necessitates the collective strategies to participate in local development planning, to adopt inclusive measures, and to take care of insiders' economic entitlements. In addition to being the gatekeepers of community interests in the urbanization process, cadres at the grassroots level adapted their self-development strategies to fit the government's dialog of beautification and modernization, and their inclusive measures attracted more new arrivals, who contributed to the prosperity of local economy.

These socioeconomic factors not only contributed to the uniqueness of Northwestern China, but also shaped the difference within that region. Despite their common institutional setting and similar development environment, the two inland villages investigated in this study differ in several respects. The decentralization of their grassroots leadership was in part determined by their proximity to the urban center of Yinchuan. Related to this, the villages varied in their capacity to mobilize villagers and economic resources to retain and maximize indigenous economic power. In Ning Cun, grassroots leadership was undermined because the village was close to the urban center. In contrast, the local collective leadership in Pan Cun was left intact and with more time to prepare for land expropriation of land. But from another perspective, the geographical proximity of Ning Cun to the urban center contributed to the village's attractiveness and added to the locals' bargaining power, which was not the case in Pan Cun. In addition, the composition of newcomers shaped the spatially bounded economic power of native residents in important ways. Newcomers either contributed to indigenous economic power by working as low-pay laborers and tenants or undermined this power by intensifying market competition and penetrating the indigenous space to maneuver as new homeowners. As shown in this study, the grassroots collectives in Ning Cun and Pan Cun implemented both cooperative and inclusive measures, but in different ways, to gain official approval, to make their villages desirable, and to defend the economic interests of the "local citizenship." Therefore, researchers should consider the great diversity of local collectives that lie behind apparently similar grassroots initiatives that promote self-help development and invest in local production, infrastructure, and amenities.

Such distinctions have far-reaching implications for the place-making outcomes and the rural–urban gap on China's new urban frontier. Given the need for investment and the prospect of rewards, local residents should invest in their indigenous economic resources with confidence in promising returns, whereas the two villages were troubled by either the uncertain commercial desirability of the place or the withering economic control over property rights. Still, the two cases under study suggest alternative scenarios other than deprivation and hopelessness for villagers in the face of urban sprawl. Backed up by collective entitlements, the grassroots self-development projects have economically empowered the native villagers, provided villages with a sense of self-determination, and helped to reduce the rural-urban gap in social and spatial terms.[16] In the newly formed communities with a residential mix of insiders and outsiders, and between people with rural and urban origins, however, the rural–urban gap might be retained and reproduced in other forms with regard to people's diverse lifestyles and socioeconomic entitlements. With the arrival of more propertied newcomers and the withering of grassroots authorities, it is yet to be examined how the space to maneuver is sustainable.

Acknowledgment

This work was supported by the Research Grants Council of Hong Kong Special Administrative Region [HKBU12406714] and the Research Committee of Hong Kong Baptist University (FRG1/13-14/059, FRG2/12-13/051 and the start-up grant).

Notes

1. The subnational divisions include 22 provinces as well as 5 autonomous regions (including Ningxia, the field site), 4 municipalities, and 2 special administrative regions. The provincial-level divisions are mostly divided into prefecture-level cities and other equivalent-level administrative units. The prefecture-level cities (including Yinchuan, the field site) are mostly divided into districts, counties, or county-level cities. At a lower level of government, districts are comprised of subdistricts or street offices, and counties have townships to govern rural areas. At the lowest level, street offices are subdivided into residential communities, and townships are composed of villages. They are respectively governed by community committees or village committees (typically divided into villagers' groups).
2. Their counterparts in the coastal areas witnessed a fever to build development zones in the 1990s, which was soon cooled down by the regulatory measures of the central government.
3. The field sites under study benefited from their favorable location near the Yellow River, but their development options were still spatially restricted by the arid climate and the scarcity of water resources.
4. The research team includes Professor Yang Shanhua, Cheng Weimin, and other colleagues at Peking University. The author joined the team as a collaborator. The team paid field visits in 2002, 2003, 2005–2007, 2009, 2010, and 2014. The multiple field visits allowed us to follow up the land development process, to detect variations in people's responses, and to compare how people memorized and interpreted social changes.
5. We got to know several key informants introduced by our local contacts in each village, who were able and willing to introduce us to other respondents (one major respondent from each household). The snowball-sampling procedure was used because it was difficult to obtain a complete sampling frame because rural communities were in a process of being replaced by residential communities with a mix of indigenous residents and outsiders. People may have changed their residences and workplaces that were not yet reflected in their official hukou status.
6. During the decollectivization of China's rural economy in the early 1980s, the term "villagers' small groups" officially replaced the Maoist term "villagers' production teams" (*dui*). However, the old term was still used by the villagers in our sample, and the term "team," which reflects their sense of ownership and belonging as expressed in daily dialog, is also adopted in this paper.
7. The greater the area of land taken away from the family, the less they paid for the house. If the household included three landholding members, the family was only required to pay half

of the construction costs (approximately 100,000 yuan). A similar villa-style house in one of the nearby suburban areas would have cost 250,000 yuan in 2005, and approximately 350,000 yuan in 2010.

8. In 2006, Team Four owned 400 mu of farmland and comprised 60 households. In 2009, it accommodated more than 80 households, and "land-loss" newcomers from the suburban areas in which land was developed were entitled to local *hukou*.

9. China's accumulated land expropriation increased very fast in early 1990s and gradually stabilized after 1994. The "zone fever" was followed by the governments' cool-down measures in the mid and late 1990s, in order to maintain a reasonable pace of land expropriation.

10. The Greater Yinchuan project was designed to expand the municipality's existing built-up areas into the surrounding countryside and establish a center of trade and logistics in Northwest China, including a market in fruits and vegetables, a plaza for the trade of construction materials, a furniture shopping mall, and an automobile showroom. By 2013, the municipality had initiated 553 "key point construction projects" (*Ningxia Daily*, 25 September 2013) and proudly claimed that it ranked 69th of 289 Chinese cities in terms of attractiveness to investors (*Yinchuan Evening Paper*, 4 September 2013).

11. Rural communities were to be transformed into urban neighborhoods, and the townships and villages lost their authority and autonomy at an unprecedented pace – their leadership would ultimately be replaced by urban street offices and residential community committees.

12. As the central government oscillated between stringent land use control and intentionally ambiguous land policies (Ho 2001), local governments were directly responsible for addressing resentment at the grassroots level (Yep 2013).

13. Local governments were reluctant to acknowledge the formal property rights of Spring Garden. Property transactions were carried out using incomplete property certificates.

14. Spring Garden residents delayed their payment of management fee, and thus, until 2005, the heating system began to function, making the community much more "livable." But Spring Garden still lacked ancillary construction. A young native respondent chose to buy commercial housing in Yinchuan to ensure that her only son would live in a district with good schools (*xuequfang*), although her Spring Garden home had been a rewarding investment.

15. Ning Cun's villagers' group was only responsible for the management of certain collective properties and utilities and was due to be replaced by urban residents' committees and commercial property management companies.

16. As villagers in Ning Cun commented, "Now urban residents wanted to buy our houses," given the increasing desire among city dwellers to move to the new urban fringe for affordable housing. Such labor and residential mobility was expected to increase the commercial desirability of the place, in which indigenous rural residents were not displaced.

References

Ahlers, Anna L., and Gunter Schubert. 2009. "Building a New Socialist Countryside – Only a Political Slogan?" *Journal of Current Chinese Affairs* 38: 35–62.

Chan, Kam Wing. 1994. *Cities with Invisible Walls: Reinterpreting Urbanization in Post-1949 China*. Hong Kong: Oxford University Press.

Chan, Kam Wing. 2010a. "A China Paradox: Migrant Labor Shortage amidst Rural Labor Supply Abundance." *Eurasian Geography and Economics* 51: 513–530.

Chan, Kam Wing. 2010b. "Fundamentals of China's Urbanization and Policy." *The China Review* 10: 63–94.

Chan, Kam Wing, and Will Buckingham. 2008. "Is China Abolishing the Hukou System?" *China Quarterly* 195: 582–606.

Chio, Jenny. 2014. *A Landscape of Travel: The Work of Tourism in Rural Ethnic China*. University of Washington Press.

Chung, Him. 2013. "The Spatial Dimension of Negotiated Power Relations and Social Justice in the Redevelopment of Villages-in-the-City in China." *Environment and Planning A* 45: 2459–2476.

Chung, Him. 2014. "Rural Transformation and the Persistence of Rurality in China." *Eurasian Geography and Economics* 54: 594–610.

Day, Alexander. 2013. *The Peasant in Postsocialist China: History, Politics, and Capitalism*. Cambridge: Cambridge University Press.

Deng, F. Frederic, and Youqin Huang. 2004. "Uneven Land Reform and Urban Sprawl in China: The Case of Beijing." *Progress in Planning* 61: 211–236.

Gaubatz, Piper. 2005. "Globalization and the Development of New Central Business Districts in Beijing, Shanghai, and Guangzhou." In *Restructuring the Chinese City: Changing Society, Economy and Space*, edited by Fulong Wu and Laurence Ma. 98–121. New York: Routledge.

Gaubatz, Piper. 2008. "Commercial Redevelopment and Regional Inequality in Urban China: Xining's Wangfujing?" *Eurasian Geography and Economics* 49: 180–199.

Hale, Matthew A. 2013. "Reconstructing the Rural: Peasant Organizations in a Chinese Movement for Alternative Development." https://digital.lib.washington.edu/researchworks/handle/1773/23389.

Han, Sunsheng, and Bo Qin. 2005. "The Cities of Western China: A Preliminary Assessment." *Eurasian Geography and Economics* 46: 386–398.

Hansen, Mette H. 2004. "The Challenge of Sipsong Panna in the Southwest: Development, Resources, and Power in a Multiethnic China." In *Governing China's Multiethnic Frontiers*, edited by Morris Rossabi, 53–83. Seattle, WA: University of Washington Press.

Hayter, Roger, and Sunsheng Han. 1998. "Reflection on China's Open Policy Towards Foreign Direct Investment." *Regional Studies* 32: 1–16.

He, Shenjing. 2007. "State-sponsored Gentrification under Market Transition." *Urban Affairs Review* 43: 171–198.

He, Shenjing, and Wu Fulong. 2007. "Socio-spatial Impacts of Property-led Redevelopment on China's Urban Neighbourhoods." *Cities* 24: 194–208.

Ho, Peter. 2001. "Who Owns China's Land? Policies, Property Rights and Deliberate Institutional Ambiguity." *The China Quarterly* 166: 394–421.

Hsing, You-tien. 2010. *The Great Urban Transformation: Politics of Land and Property in China*. New York: Oxford University Press.

Leaf, Michael. 2002. "A Tale of Two Villages: Globalization and Peri-urban Change in China and Vietnam." *Cities* 19 (1): 23–31.

Liang, Yunbin, and Yong Zhou. 1993. "Guanyu woguo chengshi kaifaqu de chubu yuanjiu [Preliminary Study on Urban Development Zones]." *Chengshi guihua* [City Planning Review] 4: 27–30.

Lin, Nan. 1995. "Local Market Socialism: Local Corporation in Action in Rural China." *Theory and Society* 24: 301–354.

Lin, George. 2010. "Understanding Land Development Problems in Globalizing China." *Eurasian Geography and Economics* 51: 80–103.

Liu, Junde. 2004. "Zhongguo Zhuanxingqi tuxian de 'xingzhengqu jingji' xianxiang fenxi [An Analysis of 'Jurisdictional Economy']. *Lilun qianyan* [Theoretical Frontiers] 10: 20–22.

Liu, Yuting, Shenjing He, Fulong Wu, and Chris Webster. 2010. "Urban Villages Under China's Rapid Urbanization: Unregulated Assets and Transitional Neighbourhoods." *Habitat International* 34: 135–144.

Looney, Kristen. E. 2012. "China's 'Building a New Socialist Countryside': The Ganzhou Model of Rural Development." 2012 Annual Meeting Paper of the American Political Science Association, Georgia Institute of Technology, Atlanta, GA, October 12–14. http://ssrn.com/abstract=2110705.

Lora-Wainwright, Anna. 2012. "Rural China in Ruins: The Rush to Urbanize China's Countryside is Opening a Moral Battleground." *Anthropology Today* 28: 8–13.

Ma, Laurence J. C. 2005. "Urban Administrative Restructuring, Changing Scale Relations and Local Economic Development in China." *Political Geography* 24: 477–497.

Matthew, Thomas and Cliff, James. 2009. "Neo Oasis: The Xinjiang Bingtuan in the Twenty-first Century." *Asian Studies Review* 33: 83–106.

Naughton, Barry. 1988. "The Third Front: Defense Industrialization in the Chinese Interior." *The China Quarterly* 115: 351–386.

Oi, Jean C. 1989. *State and Peasant in Contemporary China: The Political Economy of Village Government*. Berkeley: University of California Press.

Oi, Jean C. 1992. "Fiscal Reform and the Economic Foundations of Local State Corporatism in China." *World Politics* 45: 99–126.

Paik, Wooyeal, and Kihun Lee. 2012. "I Want to Be Expropriated!: The Politics of *xiaochanquanfang* Land Development in Suburban China." *Journal of Contemporary China* 21: 261–279.

Pannell, Clifton W. 2011. "China Gazes West: Xinjiang's Growing Rendezvous with Central Asia." *Eurasian Geography and Economics* 52: 105–118.

Pannell, Clifton W., and Philipp Schmidt. 2006. "Structural Change and Regional Disparities in Xinjiang, China." *Eurasian Geography and Economics* 47: 329–352.

Peng, Sen. 2001. *Zhongguo jingjitequ kaifaqu nianjian* [China Economic Development Zone Yearbook]. Beijing: China Financial and Economic Press.

Po, Lanchih. 2008. "Redefining Rural Collectives in China: Land Conversion and the Emergence of Rural Shareholding Co-operatives." *Urban Studies* 45: 1603–1623.

Ren, Bo. 2003. "Xin quandiyundong de molu [The End of the New Land Enclosure Movement]." *Caijing Magazine* 16: 50–55.

Schubert, Gunter, and Anna L. Ahlers. 2012. "County and Township Cadres as a Strategic Group: 'Building a New Socialist Countryside' in Three Provinces." *China Journal* 67: 67–86.

Smart, Alan, and Josephine Smart. 2001. "Local Citizenship: Welfare Reform Urban/rural Status, and Exclusion in China." *Environment and Planning* 33: 1853–1869.

Smith, Nick R. 2014. "Beyond Top-down/Bottom-up: Village Transformation on China's Urban Edge." *Cities* 41: 209–220.

Smith, Christopher J., and Katie M. G. Himmelfarb. 2007. "Restructuring Beijing's Social Space: Observations on the Olympic Games in 2008." *Eurasian Geography and Economics* 48: 543–554.

Veeck, Gregory, ed. 1991. *The Uneven Landscape: Geographic Studies in Post-reform China.* Baton Rouge, LA: Louisiana State University.

Walcott, Susan M. 2003. "Xi'an as an Inner China Development Model." *Eurasian Geography and Economics* 44: 623–640.

Walder, A. 1992. "Property Rights and Stratification in Socialist Redistributive Economics." *American Sociological Review* 57: 524–539.

Wang, Winnie Wenfei, and Cindy C. Fan. 2012. "Migrant Workers' Integration in Urban China: Experiences in Employment, Social Adaptation, and Self-identity." *Eurasian Geography and Economics* 53: 731–749.

Wu, Fulong. 2000a. "Place Promotion in Shanghai, PRC." *Cities* 17: 349–361.

Wu, Fulong. 2000b. "The Global and Local Dimensions of Place-making: Remaking Shanghai as a World City." *Urban Studies* 37: 1359–1377.

Wu, Fulong, Fangzhu Zhang, and Chris Webster. 2012. "Informality and the Development and Demolition of Urban Villages in the Chinese Peri-urban Area." *Urban Studies* 50: 1919–1934.

Xu, Feng. 2008. "Gated Communities and Migrant Enclaves: The Conundrum for Building 'Harmonious Community/ shequ'." *Journal of Contemporary China* 17: 633–651.

Yep, Ray. 2013. "Containing Land Grabs: a Misguided Response to Rural Conflicts Over Land." *Journal of Contemporary China* 22: 273–291.

Zhang, Tingwei. 2000. "Land Market Forces and Government's Role in Sprawl: The Case of China." *Cities* 17: 123–135.

Zhang, Li. 2010. *In Search of Paradise: Middle-Class Living in a Chinese Metropolis.* Ithaca, NY: Cornell University Press.

Zhang, L., Simon XB. Zhao, and J. P. Tian. 2003. "Self-help in Housing and Chengzhongcun in China's Urbanization." *International Journal of Urban and Regional Research* 27: 912–937.

Zweig, David. 2000. "The Externalities of Development: Can New Political Institutions Manage Rural Conflict?" In *Chinese Society: Change, Conflict and Resistance*, edited by Elizabeth J. Perry and Mark Selden, 120–142. London: Routledge.

Neighborhood conflicts in urban China: from consciousness of property rights to contentious actions

Qiang Fu

Department of Sociology, The University of British Columbia, Vancouver, BC, Canada

Although urban neighborhood conflicts have drawn widespread attention, their possible link with neighborhood perception has not been quantified in the existing literature. Based on a recent neighborhood-based survey of urban residents in Guangzhou, China, this study investigates the structure, determinants, and consequences of neighborhood conflicts. In particular, it finds that there is inherently a subjective dimension embedded in neighborhood conflicts such that these conflicts should be conceptualized and measured as an individual-level perception of neighborhoods. Evidence from both empirical analyses and field research revealed that consciousness of property rights had significant effects on perceived neighborhood conflicts, while both consciousness of property rights and perceived neighborhood conflicts, especially those with local and grass-roots government agencies, further contribute to the occurrence of residents' contentious actions. By situating neighborhood conflicts in the context of rules consciousness, this study brings attention to neighborhood perception shaping contentious politics in China's urban transformation.

Introduction

Neighborhood conflicts have been an integral part of urban politics for years (Tomba 2004; Bray 2005; Shi and Cai 2006; Read 2008; Hsing 2010; Fu and Lin 2014). Whereas the residential experience has been plagued by conflicts unfolding across communities, conflicts as a process of social interaction allow residents to develop neighborly connections beyond the control of the state and subsequently foster residents' engagement in contentious politics (Tomba 2004, 2005; Fu and Lin 2014). Nevertheless, few if any studies have quantified the structure, determinants, and consequences of neighborhood conflicts, although a flood of urban unrest during China's urban transformation has often been observed (Read 2008; Perry 2010; Yip and Jiang 2011). In the absence of such empirical evidence on neighborhood tensions and disputes, scholars remain puzzled about whether and how neighborhood conflicts are relevant to contentious politics in urban China.

Based on a recent neighborhood-based survey conducted in Guangzhou, this research thus addresses the following questions. First, given that existing studies largely treat neighborhood conflicts as objective, evenhanded, or homogeneous constructs, how do urban residents in Guangzhou experience a variety of neighborhood conflicts involving different entities of neighborhood governance (e.g. developers, property managers, local governments, or homeowners)? Moreover, is there a subjective dimension embedded in the structure of neighborhood conflicts, such that certain residents are more susceptible for perceiving problems taking place within their neighborhoods? Second, net of other effects, do different types of neighborhood conflicts relate to residents' engagement in contentious politics? If so, what types of neighborhood conflicts tend to be more pronounced in affecting contentious actions? Following a brief discussion on the major restructuring of neighborhood governance during China's great urban transformation, the sections below illustrate the relevance of these research questions to the existing literature in human geography, urban sociology, and political science, in particular, neighborhood perception and rules consciousness. By investigating neighborhood conflicts from a multidisciplinary perspective, scholars can gain additional insights for evaluating the multifaceted impacts of China's urban transformation.

Farewell to the workplace: transforming China's urban neighborhoods

The major restructuring of neighborhood governance derives from several consequential changes during China's urban transformation: the abolition of a workplace-based housing system, the establishment of an urban housing market, and the consolidation of grass-roots government agencies inherited from the pre-reform era. In the pre-reform era, the very existence of a housing market contradicted the socialist ideology of universal housing provision for urban residents (Szelenyi 1983; Logan, Bian, and Bian 1999). Since important urban resources and capital were mainly produced, possessed, and distributed via workplaces (work units or *danwei*) located throughout China's political hierarchy, access to urban housing was achieved by paying trivial amounts of rent to urban residents' corresponding workplaces (Wang and Murie 1996; Li 2000; Davis 2003; Huang 2004; Fu 2015a). Urban planning in the pre-reform era emphasized the link between workplace and residence when urban residents were considered more as producers in workplaces than housing consumers in residential communities. These workplace compounds that dominated urban space allowed residents from a single workplace to occupy the same residential space, which was often proximate to their workplace. This residential arrangement accommodated the needs of socialist production and urban administration (Lo 1994; Wang and Murie 2000; Huang 2003; Li 2003; Lu 2006; Fu 2015a). Moreover, the socialist principles of urban design included, among others, the integration of housing and facilities, optimal residential grouping determined by service radius (the distance between housing and location of services), and a match between the provision of community facilities and number of residents being served (Lu 2006; Fu 2015a).

In the pre-reform era, local authorities were seldom concerned about neighborhood conflicts because these conflicts were mainly mitigated, solved, or controlled by hierarchical relations within the workplace (Cai 2008b). A study on political participation and contentious politics in the pre-reform era shows that of the 757 residents interviewed in Beijing, over half reported that they had contacted workplace leaders, whereas only 4 percent had conveyed grievances to complaint bureaus in the period 1983–1988 (Shi 1997). Several factors accounted for the paramount importance of workplaces in addressing pre-reform urban conflicts. First, given that socialist housing provision was virtually a welfare policy (Saunders 1984), the property rights of both housing and residential compounds were held by workplaces rather than private households. Low-rent housing and access to a variety of community facilities (e.g. grocery stores, kindergartens, and/or bathhouses) were deemed as privileges conferred by workplace affiliation. Urban residents did not perceive themselves as stakeholders of either housing or communal space, but appealed to workplaces to solve neighborhood problems. Second, because urban residents living in the same workplace compound were both neighbors and co-workers, workplace-based networks presumably promoted neighborly interactions, mutual trust, and reconciliation within a neighborhood (Zhu, Breitung, and Li 2012; Fu et al. 2015). Third, since workplaces had been key institutions through which the Communist Party exercised its control over urban space, the threat of punishment from workplaces and political degradation also discouraged contentious actions by urban residents.

Although the pre-reform principles of urban planning should by no means be interpreted as hampering the residential experience and neighborly interactions, a widespread problem embedded in socialist housing provision was the serious underinvestment in urban housing. Being regarded as a welfare policy, local authorities lacked the incentive to expand housing provision because such welfare provision generated little revenue for local governments or workplaces and instead consumed substantial capital and resources that could have been used in industrial production (Fu and Lin 2013). In order to remove urban housing provision from the central and local budgetary sheets, piecemeal housing reforms since the 1980s have gradually allowed, embraced, and advocated an urban housing market in China (Wang and Murie 1996; Wang 2001; Li and Yi 2007). Despite housing subsidies provided to the state-sector (governments, institutes and state-owned enterprises) workers (Wang 2001), as workplaces increasingly retreat from urban housing provision, many workplace compounds are being replaced by gated communities consisting of commodity housing units purchased on the open market (Wu 2001, 2002; Davis 2003; Lu 2006). Meanwhile, property management of new commodity-housing neighborhoods is now being assumed by professional property-management companies (PMCs), which deliver services of sanitation, gardening, security, maintenance of facilities, grounds keeping, and so on by charging residents monthly property management fees (Read 2003; Fu and Lin 2014). Bearing in mind that PMCs in urban China are often established or introduced by developers who maintain reciprocal relations with local governments and their grass-roots arms, PMCs largely possess power over urban residents. This triggers neighborhood conflicts regarding exorbitant property management fees, appropriation of maintenance funds, unauthorized changes in neighborhood planning, and so on (Lin and Ho 2005; Fu and Lin 2013; Fu 2015b).

Finally, grass-roots government agencies inherited from the pre-reform era, such as street offices (*Jiedaoban*) and their executive branches, and residents' committees (*Juweihui,* also called community work stations in some Chinese cities such as Shenzhen), have not only survived but thrived in the reform era. The subject of grass-roots government agencies in China deserves further attention because they differ from self-organized (Tocquevillian) civic organizations in democratic societies or local governments. Consisting of several to a dozen civil servants and supporting staff who are either nominated or approved by higher level governments, grass-roots government agencies are empowered through a top-down effort to control territory-based resources and personnel (Bray 2006; Hsing 2010; Lee and Zhang 2013). Since grass-roots state agencies are relied on to maintain social stability, Lee and Zhang (2013) note that urban residents, especially these aggrieved ones, often directly interact with grass-roots government agencies instead of local governments at the district, county, or city level.[1]

As the move toward a market economy led to personnel and resources outside the command economy system, workplace-centered administration has been gradually shifted to territory-oriented governance; urban individuals without formal workplace affiliations, such as temporary migrants and laid-off workers, can thus be covered by territory-based governmental agencies (Wu 2002; Bray 2006). Drawing on the institutional legacy of the planned economy, the empowering of grass-roots government agencies has been deliberately employed as a strategy to restore the party's authority over urban space. For example, it is stipulated that homeowners associations, territory-based voluntary organizations elected by homeowners, must be directed, monitored, and supervised by state-sponsored residents' committees (Ministry of Construction 2009). Through demolition and urban renewal projects, a contentious process of territorialization further allows grass-roots governments to prosper by claiming authority over valuable urban space previously occupied by various workplaces (Hsing 2010).

A neighborhood of their own: understanding the causes and consequences of neighborhood conflicts

Scholars across disciplines have different yet related interpretations of the association between neighborhood conflicts and contentious politics. The effects of neighborhood conflicts on contentious actions have been well documented by existing studies (Martin 2003; Tomba 2004, 2005). Despite the demise of neighborly interactions based on previous workplace-based connections, neighborhood conflicts allow urban residents with diversified social and economic origins in newly developed neighborhoods "to overcome the architectural barriers," and have become "a very important source of socialization" (Tomba 2005, 946). More importantly, neighborly interactions triggered by conflicts allow isolated residents who barely know each other to develop social ties beyond the control of either workplaces or governments, which provide a network basis from which contentious activism flexes its muscle (Fu and Lin 2014; Fu et al. 2015). Neighborhood conflicts stimulate civic interactions in the public sphere and promote "schools of democracy," to quote Putnam (2000), which not only allow residents who share the same residential interests to build solidarity and mutual trust, but subsequently form territory-based identities. This effect of conflicts on neighborhood perception has been supported by existing studies on neighborhood governance in reform-era China (Read 2003; Shi and Cai 2006; Yip and Jiang 2011).

Although there is a dearth of literature inquiring into the reverse link from neighborhood perception to neighborhood conflicts, neighborhood perception has been widely adopted by geographers, sociologists, and epidemiologists to predict a series of neighborhood-level and person-level outcomes such as violent crimes, civic engagement, environmental protection, and depressive symptoms (Sampson, Raudenbush, and Earls 1997; Dunlap 2002; Martin 2003; Ahern and Galea 2011). Moreover, social scientists have long observed that conditions do not necessarily result in *conflicts* unless they are first perceived as *problems* by related claims makers, who then publicize their perception to stakeholders, legitimize their claim in public discourse, and finally appeal to contentious politics to address these problems.

Based on the premise that place generates both common residential experience and a powerful mobilizing discourse, Martin (2003) proposes that neighborhood identity that obscures social and economic differences among residents but stresses territory-based perception fosters, if not enables, neighborhood activism. While Martin's work echoes the discussion on neighborhood identity or sense of place from the 1980s (Smith 1984, 1985), this paradigm of place framing as place making differs from its predecessors because it suggests a causal direction from neighborhood perception to activism. More specifically, such territorialization of social perception is essentially identity politics in the sense that advocating a territory-based identity over other forms of identities is more of a social construction than an a priori neighborhood condition, which neighborhood activists deliberately project onto the residential experience for an instrumental purpose. These activists use the strategy of locating and blaming a variety of neighborhood problems that jeopardize the common residential experience, and thus should be addressed to advance residents' engagement in neighborhood affairs (Martin 2003).

With the massive transfer of housing property rights from workplaces to households, neighborhood engagement in urban China now also depends on whether residents are aware of their partitioned ownership (or condominium) within a neighborhood (Davis and Lu 2003; Fu and Lin 2014; Fu et al. 2015). Meanwhile, the observation that neighborhood activists equipped with adequate knowledge of neighborhood conditions and property rights have better understanding of neighborhood problems lends support to the reverse causality from neighborhood perception to conflicts or a subjective dimension of neighborhood conflicts. In this regard, residents who become aware that their current interests as housing consumers are fundamentally different from those of socialist producers treating workplace housing as a means of production (Tomba 2005; Zhang 2008) are more susceptible to neighborhood conflicts.

Neighborhood conflicts: from rules consciousness to contentious politics

In their seminal work on *contentious politics*, Tilly and Tarrow (2006) propose a relational account of contention and define it as making claims involving alter's interests. Contentious politics can thus be defined as "interactions in which actors make claims bearing on someone else's interests, leading to coordinated efforts on behalf of shared interests or programs, in which governments are involved as targets, initiators of claims, or third parties" (Tilly and Tarrow 2006, 4). According to these relational definitions, contentious politics takes place as long as state power is involved in dyadic claim making. Given that political institutions in China are far more ubiquitous than industrial societies and their power is often interwoven with economic activities (Oi 1997; Fu and Lin 2013), it is difficult to draw a clear border between contention and politics. Instead, their link has been emphasized in debates about the nature of contentious politics in contemporary China (Cai 2008b; Perry 2008; Li 2010).

Scholars argue that the tide of "rights talk" during popular protests, especially those conducted by a nascent middle class in urban China, embodies a bottom-up claim of democratic citizenship in an authoritarian state in which the protection of individual inalienable rights related to life, liberty, and property is both historically and presently inadequate (Tomba 2004; O'Brien and Li 2006; Zhang 2008; Li 2010). Given that rights consciousness corresponds to "the ongoing, dynamic process of constructing one's understanding of, and relationship to, the social world through use of legal conventions and discourses" (McCann 1994, 7), a growing sense of rights consciousness among citizens who are willing to assert their rights could lead to a fundamental restructuring of social and power relations in Chinese society (Li 2010).

By comparing the tradition and reality of the Chinese conception of "rights" with those in the Anglo-American tradition, Perry (2008) questions whether contentious "rights talk" should be interpreted as a signal that citizens in China are ready to embrace democracy. Different from the Jeffersonian concept of governance building on liberty and legal justice, she demonstrates that the political legitimacy of Chinese Governments has historically centered on their capacity to provide economic welfare and social security. In a society in which social citizenship (emphasizing subsistence and livelihood needs) makes more sense than either civil citizenship (asking for independent legal protections from state intrusion) or political citizenship (demanding engagement in the exercise of state power), contentious actions via state-prescribed channels in lieu of being perceived as a naturally endowed defense of civil liberty against state forces are rooted in rules consciousness rather than rights consciousness (Perry 2007, 2008). According to Perry,

> instead of indicating some novel expression of proto-democratic citizenship or state vulnerability, the continuing adherence to rules consciousness ... reflects a seasoned sensitivity on the part of ordinary Chinese to (changing yet still powerful) top-down signals emanating from the state. (2007, 21)

Rules authorized by the central authority instead of Locke's notion of *rights* are more relevant to contentious politics in China. Through the lens of a rules-consciousness framework, residents' material pursuits in contentious politics are disguised by intricate strategies advocating neighborhood engagement, such as strengthening territory-based identities or adherence to ideological slogans sanctioned by the Communist Party. Rights talk is only invoked as a strategic bulwark against unchecked local authorities. Contentious politics will not weaken, but instead strengthen the regime by replacing corrupt, unresponsive, or inept officials because rules consciousness directs resentment toward local officials instead of the overall political regime per se (Cai 2008a; Perry 2008).

The generic distinction between rights consciousness and rules consciousness (O'Brien 2001; Perry 2008; Li 2010) is articulated as follows. Rights consciousness presumes a circumscribed sphere of state sovereignty, insists that both the ruled and *the central authority* are obligated to abide by civil rights and constitutional principles, and demands both indirect and direct participation in rule making. In contrast, rules consciousness accepts laws and policies stipulated by the central authority, but remains skeptical about law- and rule-enforcement authorities at the local level, with an understanding that both the ruled and rule-enforcement authorities are equal in their relation to laws. In summary, rules consciousness demands a say in the *local* rule-enforcement process but views as granted the idea that the rule-making process is performed by the *central* authority. From the rules-consciousness perspective, neighborhood conflicts that either contradict rules stipulated by the central government or result from local and grass-roots governments' failure to satisfy basic (residential) needs are most pronounced in predicting contentious actions.

The subtle difference between rights consciousness and rules consciousness, which is implied by the existing literature, lies in the observation that the former is more relevant to subjective motives but the emphasis of the latter tends to be on instrumental goals pursued by dissidents and reformers. As clearly demonstrated by a historical account of popular protest in China (Perry 2010), China's authoritarian state is more likely to tolerate, negotiate with, or even satisfy submissive or deferential individuals making contentious claims via state-prescribed channels. Thus, the question at stake for participants in contentious politics in China is perhaps not what ideas motivate their participation, but how they can choose among various repertoires to achieve their instrumental goals while minimizing the risk of severe political punishment. At the same time, we should acknowledge that rules consciousness and rights consciousness are not contradictory, but complementary concepts (Li 2010). First, since rules consciousness and rights consciousness significantly deviate from duty consciousness assuming citizens' unconditional tolerance of unchecked authority, both concepts consist of awareness of legal orders, rights discourses, and social mobilization in the face of state intrusion. Second, both concepts are inherently heterogeneous at the individual level because individuals with diversified economic status, educational background, and social experiences tend to exhibit different perceptions of laws and legal protections when a specific issue is at stake. This similarity supports a subjective dimension in the perception of property rights and neighborhood conflicts.

Data, variables, and methods

Data

Empirical analyses were based on a recent survey conducted in 38 urban neighborhoods in Guangzhou, China. As the traditional center of the Cantonese society with strong connections abroad, Guangzhou has been at the leading edge of China's urban growth, social development, political reforms, and market transformation for centuries (Vogel 1990). In this study, urban residents were selected using a multistage stratified random sampling method. In the first stage, three primary sampling unit (PSU) strata in Guangzhou were determined by purpose of land use and population density: the inner core area, the inner suburb area, and the outer suburb area. In the second stage, street offices within each stratum were selected with reference to the total number of street offices located in each stratum and the spatial distribution of these street offices in urban Guangzhou. By the end of 2012, the inner core area, the inner suburb area, and the outer suburb area located within the border of the outer ring road of Guangzhou had 52, 45, and 42 street offices, respectively, which served as a sampling frame of street offices. In the third stage, a GIS sampling method was used to select one urban neighborhood and a list of two to eight adjoining ones (the exact number depending on the geographical sizes of these neighborhoods) within a selected street office. When interviewers failed to recruit a reasonable number of cooperative residents, an alternative neighborhood was selected from this list. Because few urban neighborhoods are located in the outer suburb area of Guangzhou, this area had the lowest number of accessible urban neighborhoods included in this study. In the fourth stage, adult residents within an urban neighborhood were recruited using an interval sampling procedure based on the residential distribution of a neighborhood. The resident's neighbors were chosen alternatively if the initial respondent refused to participate or complete the interview. The final sample used for this study consists of 1674 residents surveyed at the end of 2012.

Variables

The empirical analysis included both the presence and count of a series of neighborhood conflicts with different entities (real estate developers, PMCs, governments, and homeowners). Because PMCs and real estate developers in China can derive political power from their coalitions with grass-roots government agencies (Fu 2014), the first three categories other than conflicts among homeowners themselves were more relevant to the realm of contentious politics. Considering that contentious actions within a Chinese urban neighborhood often result from contested property rights (Shi and Cai 2006; Read 2008), an additional category of *conflicts over property* was generated by retrieving information on conflicts over private or condominium ownership. In particular, these conflicts relate to changes in nearby neighborhood planning by either developers or government agencies, ancillary facilities, building quality, the size of shared areas, temporary power and water supply, failure to fulfill contracts, failure by either developers or government agencies to issue certificates of housing or common property, appropriation by either property management companies or government agencies of revenues derived from homeowners' common property, homeowners' changes in the structure or residential purpose of housing, homeowners' abuse of common property, the allocation of property maintenance costs, and the raising and use of property maintenance funds.

Contentious actions is a dichotomous variable measuring whether a resident engaged in any of the following activities over the last 12 months: forming a petition (*shangfang*), participating in a protest, openly supporting a joint letter, reporting neighborhood problems to the media,[2] reporting neighborhood problems to relevant government agencies, filing a collective lawsuit due to housing problems, filing a claim due to administrative omission, or participating in contentious actions with homeowners from other neighborhoods. With regard to demographic covariates, this research considered a resident's sex, age, marital status (coded as one if a resident has ever married and zero otherwise), and local residency (coded as one if a resident holds local de jure residency, or *hukou*, and zero otherwise). For socioeconomic status, housing tenure status, years of schooling, occupation, affiliation with state sectors (i.e. government agencies, institutes, and state-owned enterprises), and party membership were considered. Finally, three covariates – perceived civic engagement, informal social control, and consciousness of property rights – obtained from factor analyses were employed to measure neighborhood perception (see Appendix 1 for the questions used to generate factor scores).

Method

Empirical analyses proceeded as follows. To investigate which variables correlated with neighborhood conflicts, their net associations with a series of socioeconomic, demographic, and behavioral variables were demonstrated in a partial correlation matrix. After identifying a possible net association between consciousness of property rights and neighborhood conflicts, this association was further evaluated using propensity-score models. By using logistic models to model the propensity for treatment or differential levels of the key variable having a strong net association with neighborhood conflicts, propensity-score models estimated the treatment effects (the effects of the key variable on neighborhood conflicts) by balancing all other covariates between *treated* (a treatment group with a higher propensity for the key variable) and *untreated* (a control group with a lower propensity for the key variable) groups. By comparing treatment effects among subgroups of a survey sample with balanced characteristics, yet stratified by the propensity for treatment, statisticians have shown that this method is helpful in estimating the partial effect of the key independent variable being studied, as long as the ignorability assumption (treatment assignment is independent of the outcomes conditional on observed data) is not violated (Rosenbaum and Rubin 1983; Angrist, Imbens, and Rubin 1996). Finally, logistic regression models were employed to examine factors predicting contentious actions. The findings of this research are also complemented by qualitative evidence from field research on homeowners and homeowners associations in Guangzhou from 2009 to 2012.

Results

Table 1 shows the frequency distributions of neighborhood conflicts adjusted by size of neighborhood. Less than 30 percent of residents reported various sorts of conflicts with developers. In particular, conflicts related to changes in neighborhood planning and ancillary facilities were most frequent. Due to the considerable amount of common property at stake, these two types of neighborhood issues, especially in a sizable neighborhood in China, can often bring lucrative benefits to developers at the cost of homeowners (Yip and Jiang 2011; Fu and Lin 2014). Conflicts with PMCs (43.61 percent) were the most frequently reported category of conflicts. Specifically, conflicts related to parking, safety, and property management fees were most prevalent. Among all categories of conflicts, fewer residents (17.62 percent) reported conflicts with local and grass-roots governments. Of this type of conflict, changes in nearby neighborhood planning were most frequently reported. A substantial share (42.23 percent) of residents reported conflicts and disputes among homeowners. Falling objects and illegal possession of animals (pets) were most frequently reported. If conflicts related to property are treated as a separate category, 43.37 percent of residents reported neighborhood conflicts over private or common property. Finally, 15.17 percent of residents had engaged in various contentious actions. Contacting government agencies, contacting the media, participating in a protest, and forming a petition were the most prevalent forms of contentious actions adopted by urban residents.

In terms of demographic covariates, less than one-half (44.32 percent) of the interviewees were male. Their average age was 44.85 years. The majority of interviewees were married and local residents. With regard to housing tenure status, about one-fifth of the interviewees were tenants, while the majority of homeowners lived in commodity housing. For socioeconomic status, the average years of schooling were 10.21 years, in the middle school to high school range. Residents holding administrative, professional, or clerical jobs accounted for 14.70, 13.86, and 6.75 percent of the sample population, respectively. Retired interviewees composed 28.26 percent of the respondents. About one-fifth of the residents worked in state sectors or held Communist Party membership (Table 2).

Based on the results from factor analyses of neighborhood perception, consciousness of property rights is denoted by a factor score summarizing residents' responses to six questions about their familiarity with a series of neighborhood-related laws, property, and revenues (see Appendix 1 for details). As shown in Table 3, responses to each of these six questions had high factor loadings (ranging from 0.8127 to 0.8802) on the factor of consciousness of property rights, which explains more than 70 percent of total variances of responses to the six questions. Two additional factors, perceived civic engagement and informal social control, were also generated by factor analyses of responses to seven questions (see Appendix 1) to capture the influence of a resident's subjective evaluation of his/her relations with a neighborhood (Sampson, Raudenbush, and Earls 1997; Sampson, Morenoff, and Earls 1999). Questions about perceived civic engagement and informal social control were pooled together for factor analyses due to theoretical and operational links between the two dimensions of neighborhood perception (Sampson, Morenoff, and Earls 1999; Forrest and Kearns 2001). Whereas responses to the four questions relevant to civic and neighborhood engagement had high factor loadings (ranging from 0.6413 to 0.8564) on the factor of perceived civic engagement, responses to the remaining three questions about crime and social control had high loadings (ranging from 0.7265 to 0.8599) on the factor of informal social control. The two factors together explain 67.06 percent of total variances of responses to the seven questions.

Partial correlation coefficients between neighborhood conflicts and other covariates are shown in Table 4. Net of other effects, consciousness of property rights was strongly

Table 1. The frequency distribution of perceived neighborhood conflicts and contentious actions.[a]

Categories of neighborhood conflicts	Specific neighborhood conflicts	Percentage (%)
Conflicts with developers 28.38%	Changes in neighborhood planning	14.90
	Ancillary facilities	13.69
	Quality of building	10.50
	Size of shared areas	8.36
	Temporary power and water supply	8.34
	Failure to fulfill contracts	5.84
	Failure to issue property certificates	3.61
Conflicts with PMCs 43.61%	Parking	22.99
	Safety	20.26
	Property management fees	19.25
	Sanitation	15.52
	Appropriation of revenues from homeowners' common property	12.48
	Water and power supply	9.44
	Conflicts with homeowners associations	8.04
	Homeowners harassed by security guards	5.12
Conflicts with local/grassroots governments 17.62%	Changes in neighborhood planning	8.18
	Changes in nearby existing urban planning	
	Violations to partitioned ownership	
	Water and power supply	7.15
	Intervention in the election of homeowners	3.73
	associations	3.69
	Intervention in the operation of homeowners	3.30
	associations	3.21
	Appropriation of revenues from homeowners' common property	2.97
	Housing property certificates	2.96
	Property certificates of common property	2.79
Conflicts among residents 42.23%	Falling objects	31.73
	Illegal possession of animals	21.56
	Abuse of homeowners' common property	12.35
	Allocation of property maintenance costs	12.35
	Disputes over the election of homeowners associations	8.47
	Raising and use of property maintenance funds	7.49
	Changes in the structure or residential purpose of housing	6.14
	Conflicts within a homeowners association	5.07
	Conflicts between homeowners associations in the same neighborhood	2.78

(*Continued*)

Table 1. (*Continued*).

Categories of neighborhood conflicts	Specific neighborhood conflicts	Percentage (%)
Residents' contentious actions (15.17%)	*Specific contentious actions*	
	Reporting neighborhood problems to government agencies	4.52
	Reporting neighborhood problems to media	4.04
	Participating in a protest	3.62
	Creating a petition (*Shangfang*)	3.26
	Openly supporting a joint letter	2.72
	Filing a collective lawsuit due to housing problems	2.50
	Filing a claim against administrative omission	1.72
	Participating in contentious actions with homeowners from other neighborhoods	1.35

[a]The frequency distribution has been weighted according to sampling design.

Table 2. Descriptive statistics of urban residents in Guangzhou, China ($N = 1674$).

	Mean	Standard deviation
Male	44.32%	
Age	44.85	15.18
Married	81.90%	
Local resident	71.09%	
Homeownership		
Tenants	20.25%	
Homeowners of commodity housing	70.07%	
Homeowners of other types of housing	9.68%	
Years of schooling	10.21	2.95
Occupation		
Administrative	14.70%	
Professional	13.86%	
Clerical staff	6.75%	
Service, manual labor, etc.	36.43%	
Retired	28.26%	
Working in state sectors	20.49%	
Party member	21.21%	

associated with both the presence and number of neighborhood conflicts, while residents showing higher levels of perceived civic engagement were more likely to report various neighborhood conflicts. Homeowners were more alert to conflicts over property rights or with property management companies. Because the results in Table 4 reveal that the association between consciousness of property rights and perceived neighborhood conflicts is most pronounced, the effect of consciousness of property rights on neighborhood conflicts was further tested using propensity score models by dichotomizing consciousness of property rights into high (treatment group) and low (control group) levels. The results in Table 5 show that, among residents with balanced characteristics, those with high levels of consciousness of property rights reported significantly more neighborhood conflicts than those with low levels. Evidence from field research also

Table 3. Results from factor analyses of consciousness of property rights, informal social control, and perceived civic engagement.

	Rotated factor loadings Consciousness of property rights	Rotated factor loadings Perceived civic engagement	Rotated factor loadings Informal social control
D1 Maintenance fund	0.8501		
D2 Collective revenues	0.8802		
D3 Partitioned ownership	0.8468		
D4 Finances of local PMC	0.8127		
D5 Expenditures of collective revenues	0.8134		
D6 Laws and regulations	0.8503		
Eigenvalue	4.259		
Total proportion of variance explained	**70.99%**		
E1 Solve residential problem?		**0.6413**	0.3271
E2 Get involved with others?		**0.6844**	0.3526
E3 Spend your time?		**0.8564**	0.1816
E4 Spend your money?		**0.8393**	0.1721
E5 Stop vandalism?		0.1852	**0.814**
E6 Preserve neighborhood environment?		0.2041	**0.8599**
E7 Intervene in others' quarrels and fights?		0.2778	**0.7265**
Eigenvalue		3.632	1.062
Proportions of variance explained by each factor		31.77%	35.30%
Total proportion of variance explained		**67.06%**	

lends support to the observation that residents with higher consciousness of property rights are more likely to perceive neighborhood conflicts than others. Neighborhood activists (e.g. members of homeowners associations) who had a better understanding of laws and regulations related to neighborhood governance reported that they could readily identify violations of homeowners' common property even if they had moved to an entirely new neighborhood. However, there was a general lack of consciousness of property rights among most residents, who failed to recognize that their interests as housing consumers were different from those as producers previously living in a workplace compound. As suggested by a former director of a homeowners association,

> [Residents believe that] these rights [of common property] are not possessed by any owners. It is yet to be determined who the owner is. They do not know about it.... This is [in fact] not property owned by state. [It is the] private or common property of residents.

These differential levels of consciousness of property rights in turn have implications for the perception of neighborhood conflicts. For example, a neighborhood activist, who diligently engaged in establishing a homeowners association, described the differences in how residents perceived neighborhood conflicts.

Table 4. Partial correlation between perceived neighborhood conflicts and consciousness of property rights.

	Perceived conflicts over property		Perceived conflicts with developers		Perceived conflicts with property management companies		Perceived conflicts with local & grassroots governments		Perceived conflicts among homeowners	
	Presence	Number of events	Presence	Number of events	Presence	Number of events	Presence	Number of events	Presence	Number of events
Consciousness of property rights	0.1401***	0.1856***	0.1047***	0.1471***	0.0892***	0.1408***	0.1117***	0.1321***	0.1354***	0.1778***
Male	-0.0160	0.0279$^\Psi$	-0.0130	0.0192	0.0143	0.0201	0.0131	0.0288	0.0154	0.0318
Age	-0.0280	-0.0423$^\Psi$	-0.0035	-0.0327	-0.0375	-0.0374	-0.0434$^\Psi$	-0.0472$^\Psi$	-0.0328	-0.0264
Age squared	0.0192	0.0398	-0.0069	0.0292	0.0347	0.0312	0.0395	0.0458$^\Psi$	0.0349	0.0276
Married	0.0088	-0.0502*	-0.0100	-0.0314	-0.0376	-0.0320	-0.0324	-0.0544*	-0.0257	-0.0339
Local resident	0.0356	0.0096	0.0169	-0.0004	0.0338	0.0006	0.0243	-0.0005	0.0523*	0.0239
Homeowners of commodity housing	0.0494*	0.0139	0.0231	0.0063	0.0640**	0.0368	0.0239	0.0079	0.0302	0.0195
Homeowners of other types of housing	0.0769**	0.0402	0.0495*	0.0270	0.0918***	0.0685**	0.0450$^\Psi$	0.0359	0.0264	0.0245
Years of schooling	0.0311	0.0344	0.0574*	0.0327	0.0426$^\Psi$	0.0457$^\Psi$	0.0280	0.0148	-0.0196	0.0031
Administrative	-0.0144	-0.0179	-0.0174	-0.0327	0.0155	0.0136	-0.0256	-0.0196	0.0002	0.0165
Professional	-0.0119	0.0075	-0.0378	-0.0296	0.0106	0.0276	-0.0002	0.0251	0.0197	0.0304
Clerical staff	-0.0251	-0.0105	-0.0029	-0.0068	0.0081	-0.0010	-0.0180	-0.0255	-0.0046	-0.0067
Retired	0.0012	-0.0174	0.0246	-0.0046	-0.0245	-0.0143	-0.0242	-0.0269	-0.0328	-0.0287
Working in state sectors	0.0220	-0.0094	0.0232	0.0113	-0.0163	-0.0084	-0.0418$^\Psi$	-0.0408$^\Psi$	0.0101	-0.0043
Party member	0.0154	-0.0161	0.0064	-0.0106	0.0075	-0.0143	-0.0100	-0.0085	0.0177	0.0015
Perceived civic engagement	0.0423$^\Psi$	0.0396	0.0442$^\Psi$	0.0264	0.0401	0.0503*	0.0626*	0.0287	0.0202	0.0500*
Informal social control	0.0240	-0.0166	-0.0045	-0.0128	0.0120	0.0075	-0.0276	-0.0401	0.0297	-0.0068

$^\Psi p < .10$; $^*p < .05$; $^{**}p < .01$; $^{***}p < .001$ (two-tailed tests).

Table 5. Treatment effects of consciousness of property rights on counts of neighborhood conflicts estimated from propensity-score models.

	Perceived conflicts over partitioned ownership		Perceived conflicts with developers		Perceived conflicts with property management companies		Perceived conflicts with local and grass-roots governments		Perceived conflicts among homeowners	
	Difference in ATT[a]	Standard error	Difference in ATT[a]	Standard error	Difference in ATT[a]	Standard error	Difference in ATT[a]	Standard error	Difference in ATT[a]	Standard error
Consciousness of property rights	1.17***	0.18	0.43***	0.09	0.77***	0.12	0.39***	0.09	0.62***	0.12

[a]ATT = Average treatment effect.
$^{\Psi}p < .10$; $*p < .05$; $**p < .01$; $***p < .001$ (two-tailed tests).

> There were about seven or eight persons [who tried to initiate a homeowners association in this neighborhood] at the beginning. We ourselves [who tried to establish a homeowners association] understood that the property management company must be supervised. We told others that there were some problems in our neighborhood. They [other residents] also felt that there should be some issues ... but only talked about phenomena. What caused these phenomena? We did not spell them out explicitly though we understood them.... Many residents did not provide their signatures [to support the establishment of their home-owners association], they just did not understand.

As suggested by this interviewee, residents with better knowledge of property rights also had a better understanding of neighborhood problems, which in turn affected contentious actions. To further evaluate the effects of consciousness of property rights and neighborhood conflicts on contentious actions, logistic regressions were employed to investigate the determinants of contentious actions.

As expected, the results from Model 1 in Table 6 suggest that local residents and homeowners are more likely to engage in contentious actions as compared with migrants and tenants, respectively. However, the higher odds ratio of homeownership is explained by both occupational status and neighborhood perception, where perceived civic engagement and informal social control are all positively and significantly associated with the odds of contentious actions (see Model 2). Furthermore, consciousness of property rights is strongly and positively associated with contentious actions across models. In fact, leaders of urban voluntary associations have already employed the strategy of informing residents about their property rights in order to elicit civic engagement and possible contentious actions. One director of a homeowners association described how they collected enough signatures for a petition within a short time.

> We finished within one week and collected signatures from more than 5000 households. When you visit [each household in this neighborhood] you need to explain why we need their signatures and [these reasons] are listed very explicitly in [petition] forms. You cannot ask for their signatures without informing them what will happen next. Who would dare to sign [without knowing the reasons]? You need to illustrate your reasoning before collecting his[/her] signature.

According to this interviewee, residents engage in contentious actions only after they become aware of their rights and the consequences of the petition, which lends support to the rules-consciousness perspective.

The negative effects of administrative or clerical jobs on contentious actions are in line with the conceptual framework proposed by Fu and Lin (2014) in explaining neighborhood engagement in China: hidden costs such as political pressure from workplaces discourage employees' participation in contentious actions. Net of other effects, each category of neighborhood conflict also has a significant and positive effect on contentious actions (from Model 3 to Model 7); yet, the significant effect of conflicts over property is explained by conflicts with different entities in neighborhood governance when all types of conflicts are considered in Model 8. These findings further emphasize neighborhood power relations among different entities in shaping conflicts (Fu 2014; Fu et al. 2015).

Furthermore, the strong effect of conflicts with local and grass-roots governments on contentious actions (odds ratio $= e^{0.7710} = 2.1619$) also lends support to the rules-consciousness perspective if we recall that rules consciousness is associated with "skepticism toward local rule-enforcement authorities" (Li 2010, 59). Considering the

Table 6. Logistic regressions of contentious actions on perceived neighborhood conflicts.[a]

	Model 1 Coefficient	Model 2 Coefficient	Model 3 Coefficient	Model 4 Coefficient	Model 5 Coefficient	Model 6 Coefficient	Model 7 Coefficient	Model 8 Coefficient
Male	0.0962	-0.1083	-0.1031	-0.1137	-0.1355	-0.1310	-0.1318	-0.1540
Age	-0.0002	-0.0157	-0.0111	-0.0190	-0.0092	-0.0052	-0.0071	-0.0035
Age squared	-0.0001	0.0001	0.0000	0.0001	0.0000	0.0000	0.0000	-0.0001
Married	-0.1888	-0.3648^Ψ	-0.4086^Ψ	-0.3807^Ψ	-0.3215	-0.3592^Ψ	-0.3282	-0.3344
Local resident	0.7496***	0.6158**	0.5904**	0.6184**	0.5891**	0.6373**	0.5551***	0.5921**
Homeownership (tenants as the reference)								
Homeowners of commodity housing	0.6948**	-0.1906	-0.3251	-0.2464	-0.3617	-0.2679	-0.2914	-0.3587
Homeowners of other types of housing	0.6293^Ψ	-0.2483	-0.4470	-0.3771	-0.4623	-0.3732	-0.3432	-0.4844
Years of schooling		-0.0199	-0.0304	-0.0357	-0.0317	-0.0302	-0.0183	-0.0393
Occupation (manual labor and others as the reference)								
Administrative		-0.5578*	-0.5735*	-0.5462*	-0.6132*	-0.5464*	-0.6014*	-0.5880*
Professional		-0.2325	-0.2477	-0.1855	-0.2663	-0.2523	-0.2928	-0.2589
Clerical staff		-0.6981*	-0.6738*	-0.7406*	-0.7248*	-0.6555*	-0.7514*	-0.7252*
Retired		0.0582	0.0550	0.0170	0.1061	0.1171	0.1097	0.1033
Working in state sectors		0.1275	0.1233	0.1122	0.1640	0.2036	0.1376	0.1915
Party member		0.1610	0.1494	0.1647	0.1731	0.2028	0.1516	0.2005
Consciousness of property rights		0.6270***	0.5796***	0.5932***	0.6173***	0.5860***	0.5862***	0.5618***
Perceived civic engagement		0.3308***	0.3179***	0.3204***	0.3138***	0.2928***	0.3245***	0.2943***
Informal social control		0.2379**	0.2385**	0.2600**	0.2472**	0.2765**	0.2401**	0.2818**
Perceived neighborhood conflicts								
Over partitioned ownership			1.0897***					0.0535
With developers				1.0303***				0.4511*
With property management companies					0.9890***			0.3493^Ψ
With local and governments						1.3122***		0.7710***
Among homeowners							0.9790***	0.3962*
Constant	-2.5574	-1.0484	-1.4798^Ψ	-1.1633	-1.4615^Ψ	-1.4885^Ψ	-1.6238^Ψ	-1.7794**
R²	0.0291	0.1055	0.1421	0.1387	0.1363	0.1505	0.1360	0.1708
Log pseudo likelihood	-691.8878	-637.4171	-611.3918	-613.8124	-615.4816	-605.3865	-615.7267	-590.9007

[a]To account for heteroscedasticity associated with unbalanced sample sizes across different neighborhoods, statistical inference is based on robust standard errors.
$^Ψ p < .10$; $^*p < .05$; $^{**}p < .01$; $^{***}p < .001$ (two-tailed tests).

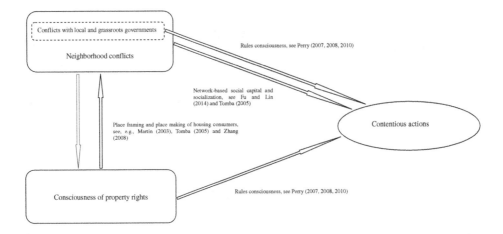

Figure 1.   A conceptual framework of neighborhood conflicts and contentious actions in Urban China.
Note: The dotted arrow from neighborhood conflicts to consciousness of property rights is not specifically addressed in the current study but is demonstrated in the existing literature (Read 2003; Shi and Cai 2006; Yip and Jiang 2011).

strong link between consciousness of property rights and perceived neighborhood conflicts (see Tables 4 and 5), these two factors also work together in affecting contentious actions. As suggested by one neighborhood activist,

> many Chinese, including homeowners from urban neighborhoods … do not care [about the neighborhood] if nothing happens. (**Interviewer:** what do you mean by 'nothing happens'?) That is, when their rights are not violated. Homeowners begin to care actively after they are hampered…. Most of us try to establish homeowners associations only after our rights are violated so that we must then protect our rights. In my neighborhood it was due to temporary water and power supply and failure in obtaining housing property certificates.This interviewee suggested a direct link between neighborhood conflicts and collective actions, which is consistent with Tomba's (2005, 946) observation that "conflicts emerging from common interests help to overcome the architectural barriers and become a very important source of socialization".

Yet, the significant marginal effect of consciousness of property rights on contentious actions (as shown in Model 2–8 at Table 6) should not be de-emphasized. Another experienced neighborhood activist pointed out that consciousness of property rights had influenced their collective actions.

> People did not know how to deal with it when homeowners' rights were violated. All they knew was fundraising and filing a lawsuit, but they did not know what the next step should be after the money was raised. … After we [neighborhood activists] arrived, we taught them how to proceed based on protocols of homeowners associations, what types of rights you were entitled to according to laws and regulations, and what you should do.

These neighborhood activists and directors of homeowners associations agreed that most urban homeowners were generally unaware of their property rights, which led to civic disengagement. The term "educating homeowners" was mentioned repeatedly during interviews.

Conclusions and discussion

Despite the growing attention to neighborhood conflicts in urban China, few if any studies have quantified the structure and distribution of neighborhood conflicts. Students of urban transformation cannot fully understand how urban space is contested unless the link among neighborhood perception, neighborhood conflicts, and contentious politics is empirically tested. Using a recent neighborhood-based survey in Guangzhou, China, this study investigates the structure, determinants, and consequences of neighborhood conflicts. It finds that there is inherently a subjective dimension embedded in neighborhood conflicts, such that these conflicts are contingent on an individual's consciousness of property rights. Evidence from both empirical analyses (logistic regression and propensity-score models) and field research revealed that consciousness of property rights had significant effects on perceived neighborhood conflicts, while both consciousness of property rights and perceived neighborhood conflicts further contribute to the occurrence of residents' contentious actions. Among all types of neighborhood conflicts, conflicts related to local and grass-roots governments are most pronounced in predicting contentious actions. These results further our understanding of neighborhood conflicts and determinants of contentious actions over urban space.

The conceptual framework and empirical findings of this study are summarized in Figure 1. This study begins with investigating whether certain residents are more susceptible to perceive neighborhood issues as conflicts. Instead of conceptualizing neighborhood conflicts as objective, homogeneous, and evenhanded, a subjective dimension is found to be embedded in neighborhood conflicts because consciousness of property rights also predicts an individual's perception of neighborhood conflicts. This subjective aspect can result in a heterogeneous mobilization process of contentious actions; whereas residents with higher levels of consciousness of property rights may perceive more neighborhood conflicts and advocate neighborhood activism, others with lower levels could stay behind as free riders. In reform-era urban China, contentious politics is in part contingent on whether propertied households recognize their current property rights and interests as housing consumers (Tomba 2004, 2005; Zhang 2008). This finding is in line with the paradigm of place framing as place making that argues that territory-based identity and knowledge can ultimately affect neighborhood engagement (Martin 2003), mediated by residents' perception of neighborhood problems and conflicts. The sense of a neighborhood worth defending and protecting is contingent upon their knowledge of property rights, informing residents of who they are, what their role in neighborhood governance should be, and why they become stakeholders of their neighborhood. Combined with the effect of neighborhood conflicts on consciousness of property rights documented in existing studies (Read 2003; Shi and Cai 2006; Yip and Jiang 2011), these two factors can perpetuate and magnify themselves in a reciprocal way, and mutually enhance contentious actions in urban China.

Neighborhood conflicts are strongly associated with the occurrence of contentious actions. They enable residents to socialize beyond architectural barriers, develop social ties in the communal space, and form a relational basis for the exchange of information, the sharing of resources, and the building of territory-based identities (Tomba 2005; Fu and Lin 2014; Fu et al. 2015). The presence and accumulation of conflicts thus facilitate contentious actions. Furthermore, of all the categories of neighborhood conflicts investigated, residential conflicts with local and grass-roots governments are most likely to be translated into contentious actions, which provide important empirical support for the rules-consciousness perspective. This finding echoes Perry's argument that social security and economic well-being serve as the backbone of China's political legitimacy. Contentious actions are prone to erupt when local governments can no longer live up to people's expectation of the government as a provider of economic prosperity or the guardian of social security, or fail to adhere to rules stipulated by the central authority (Perry 2007, 2008). Moreover, findings from the current research do suggest that contentious actions result from the synergy between neighborhood conflicts and consciousness of property rights, while each of these two factors can have a strong and independent effect on contentious actions net of other demographic, socioeconomic, and behavioral covariates. The partial, strong, and positive effect of consciousness of property rights also lends support to the rules-consciousness framework (Perry 2007, 2008), which highlights (the boundary of) institutional protocols and state regulations in shaping contentious actions. China's urban transformation not only generates private and common properties in socialist cities, but also introduces a crucial new set of rules to be inscribed in urban space for its sustainable growth. Once sanctioned by the central government, these nascent rules provide a yardstick against which the performance of local and grass-roots government agencies is to be judged, if not a basis for building civil and political citizenship.

Although this exploratory study provides useful information on neighborhood conflicts and contentious politics in urban China, several aspects of this research deserve readers' attention. First and foremost, the conclusions of this research should by no means be considered to be contradicting the rights-consciousness framework. As discussed earlier, rules consciousness and rights consciousness are two complementary concepts with different emphases. A thorough evaluation of the two notions in contentious politics requires more extensive use of qualitative and quantitative evidence pertaining to attitudes toward the central authority, local and grass-roots governments, and legal protections, which warrants further investigation of contentious politics in China. Second, although our survey research team made considerable efforts to collect information on neighborhood conflicts, the list of neighborhood conflicts provided in this study should not be regarded as an exhaustive list. Third, due to different governance modes, levels of marketization, and contexts of urban housing markets across China (Fu and Lin 2013; Fu, Zhu, and Ren 2015), the structure and prevalence of neighborhood conflicts and residential outcomes are not representative of those taking place in other Chinese cities.

Acknowledgments
The author is grateful to Kam Wing Chan, Elizabeth J. Perry, Si-ming Li, and anonymous reviewers for their valuable comments and suggestions

Disclosure statement
No potential conflict of interest was reported by the author.

Funding
This work was supported by the Hong Kong Research Grant Council [grant number HKBU245511], Lincoln Institute of Land Policy (The 2009–2010 and 2011–2012 China Program Fellowship), and the Asian Pacific Studies Institute at Duke University (the 2009 and 2011 Summer Research Fellowships).

Notes

1. Due to sensitivity of measuring contentious actions involving Chinese Government agencies, we did not distinguish grass-roots government agencies from local governments in questions about contentious actions; yet, their differences should be acknowledged.
2. Actions pertaining to the media and the legal system are considered contentious actions because state authorities in China strongly intervene, if not tightly control, these actions.

References

Ahern, Jennifer, and Sandro Galea. 2011. "Collective Efficacy and Major Depression in Urban Neighborhoods." *American Journal of Epidemiology* 173: 1453–1462.

Angrist, Joshua D., Guido W. Imbens, and Donald B. Rubin. 1996. "Identification of Causal Effects Using Instrumental Variables." *Journal of the American Statistical Association* 91 (434): 444–455.

Bray, David. 2005. *Social Space and Governance in Urban China: The Danwei System from Origins to Reform*. Palo Alto, CA: Stanford University Press.

Bray, David. 2006. "Building 'Community': New Strategies of Governance in Urban China." *Economy and Society* 35: 530–549.

Cai, Yongshun. 2008a. "Power Structure and Regime Resilience: Contentious Politics in China." *British Journal of Political Science* 38: 411–432.

Cai, Yongshun. 2008b. "Social Conflicts and Modes of Action in China." *The China Journal* 59: 89–109.

Davis, Deborah S. 2003. "From Welfare Benefit to Capitalized Asset: The Re-commodification of Residential Space in Urban China." In *Housing and Social Change: East-West Perspectives*, edited by Ray Forrest and James Lee, 183–196. London: Routledge.

Davis, Deborah S., and Hanlong Lu. 2003. "Property in Transition: Conflicts over Ownership in Post-socialist Shanghai." *European Journal of Sociology* 44: 77–99.

Dunlap, Riley E. 2002. "Environmental Sociology." In *Handbook of Environmental Psychology*, edited by Robert B. Bechtel and Arza Churchman, 160–171. New York: John Wiley & Sons.

Forrest, Ray, and Ade Kearns. 2001. "Social Cohesion, Social Capital and the Neighbourhood." *Urban Studies* 38: 2125–2143.

Fu, Qiang. 2014. "The Contentious Democracy: Homeowners Associations in China through the Lens of Civil Society." In *Housing Inequalities in Chinese Cities*, edited by Youqin Huang and Si-ming Li, 201–216. New York: Routlege.

Fu, Qiang. 2015a. "The Persistence of Power despite the Changing Meaning of Homeownership: An Age-period-Cohort Analysis of Urban Housing Tenure in China, 1989-2011." *Urban Studies*. doi:10.1177/0042098015571240.

Fu, Qiang. 2015b. "When Fiscal Recentralization Meets Urban Reforms: Prefectural Land Finance and Its Association with Access to Housing in Urban China." *Urban Studies* 52: 1791–1809. doi:10.1177/0042098014552760.

Fu, Qiang, and Nan Lin. 2013. "Local State Marketism: An Institutional Analysis of China's Urban Housing and Land Market." *Chinese Sociological Review* 46: 3–24.

Fu, Qiang, and Nan Lin. 2014. "The Weaknesses of Civic Territorial Organizations: Civic Engagement and Homeowners Associations in Urban China." *International Journal of Urban and Regional Research* 38: 2309–2327.

Fu, Qiang, Shenjing He, Yushu Zhu, Si-ming Li, Yanling He, Huoning Zhou, and Nan Lin. 2015. "Toward a Relational Account of Neighborhood Governance: Territory-based Networks and Residential Outcomes in Urban China." *American Behavioral Scientist* 59: 992–1006. doi:10.1177/0002764215580610.

Fu, Qiang, Yushu Zhu, and Qiang Ren. 2015. "The Downside of Marketization: A Multilevel Analysis of Housing Tenure and Types in Reform-Era Urban China." *Social Science Research* 49: 126–140.

Hsing, You-tien. 2010. *The Great Urban Transformation*. Oxford: Oxford University Press.

Huang, Youqin. 2003. "A Room of One's Own: Housing Consumption and Residential Crowding in Transitional Urban China." *Environment and Planning A* 35: 591–614.

Huang, Youqin. 2004. "Housing Markets, Government Behaviors, and Housing Choice: A Case Study of Three Cities in China." *Environment and Planning A* 36: 45–68.

Lee, Ching Kwan, and Yonghong Zhang. 2013. "The Power of Instability: Unraveling the Microfoundations of Bargained Authoritarianism in China." *American Journal of Sociology* 118: 1475–1508.

Li, Si-ming. 2000. "Housing Consumption in Urban China: A Comparative Study of Beijing and Guangzhou." *Environment and Planning A* 32: 1115–1134.

Li, Si-ming. 2003. "Housing Tenure and Residential Mobility in Urban China: A Study of Commodity Housing Development in Beijing and Guangzhou." *Urban Affairs Review* 38: 510–534.

Li, Lianjiang. 2010. "Rights Consciousness and Rules Consciousness in Contemporary China." *The China Journal* 64: 47–68.

Li, Si-ming, and Zheng Yi. 2007. "The Road to Homeownership under Market Transition: Beijing, 1980-2001." *Urban Affairs Review* 42: 342–368.

Lin, George C. S., and Samuel P. S. Ho. 2005. "The State, Land System, and Land Development Processes in Contemporary China." *Annals of the Association of American Geographers* 95: 411–436.

Lo, Chor Pang. 1994. "Economic Reforms and Socialist City Structure: A Case Study of Guangzhou, China." *Urban Geography* 15: 128–149.

Logan, John R., Yanjie Bian, and Fuqin Bian. 1999. "Housing Inequality in Urban China in the 1990s." *International Journal of Urban and Regional Research* 23: 7–25.

Lu, Duanfang. 2006. *Remaking Chinese Urban Form: Modernity, Scarcity and Space, 1949–2005*. New York: Taylor & Francis.

Martin, Deborah G. 2003. "'Place-framing' as Place-making: Constituting a Neighborhood for Organizing and Activism." *Annals of the Association of American Geographers* 93: 730–750.

McCann, Michael W. 1994. *Rights at Work: Pay Equity Reform and the Politics of Legal Mobilization*. Chicago, IL: University of Chicago Press.

Ministry of Construction. 2009. Yezhu Dahui He Yezhu Weiyuanhui Zhidao Guize (Guidelines for Homeowners Assemblies and Homeowners Committees). Beijing: Author.

O'Brien, Kevin J. 2001. "Villagers, Elections, and Citizenship in Contemporary China." *Modern China* 27: 407–435.

O'Brien, Kevin J., and Lianjiang Li. 2006. *Rightful Resistance in Rural China*. Cambridge: Cambridge University Press.

Oi, Jean C. 1997. "The Evolution of Local State Corporatism." In *Zouping in Transition: The Process of Reform in Rural North China*, edited by Andrew G. Walder, 35–61. Cambridge: Harvard University Press.

Perry, Elizabeth J. 2007. "Studying Chinese Politics: Farewell to Revolution?" *The China Journal* 57: 1–22.

Perry, Elizabeth J. 2008. "Chinese Conceptions of 'Rights': From Mencius to Mao – and Now." *Perspectives on Politics* 6: 37–50.

Perry, Elizabeth J. 2010. "Popular Protest: Playing by the Rules." In *China Today, China Tomorrow: Domestic Politics, Economy, and Society*, edited by Joseph Fewsmith, 11–28. Lanham, MD: Rowman and Littlefield.

Putnam, Robert D. 2000. *Bowling Alone: The Collapse and Revival of American Community*. New York: Simon & Schuster.

Read, Benjamin L. 2003. "Democratizing the Neighbourhood? New Private Housing and Home-owner Self-organization in Urban China." *The China Journal* 49: 31–59.

Read, Benjamin L. 2008. "Assessing Variation in Civil Society Organizations: China's Homeowner Associations in Comparative Perspective." *Comparative Political Studies* 41: 1240–1265.

Rosenbaum, P. R., and D. B. Rubin. 1983. "The Central Role of the Propensity Score in Observational Studies for Causal Effects." *Biometrika* 70: 41–55.

Sampson, Robert J., Stephen W. Raudenbush, and Felton Earls. 1997. "Neighborhoods and Violent Crime: A Multilevel Study of Collective Efficacy." *Science* 277 (5328): 918–924.

Sampson, Robert J., Jeffrey D. Morenoff, and Felton Earls. 1999. "Beyond Social Capital: Spatial Dynamics of Collective Efficacy for Children." *American Sociological Review* 64: 633–660.

Saunders, Peter. 1984. "Beyond Housing Classes: The Sociological Significance of Private Property Rights in Means of Consumption." *International Journal of Urban and Regional Research* 8: 202–227.

Shi, Tianjian. 1997. *Political Participation in Beijing*. Cambridge: Cambridge University Press.

Shi, Fayong, and Yongshun Cai. 2006. "Disaggregating the State: Networks and Collective Resistance in Shanghai." *The China Quarterly* 186: 314–332.

Smith, Rebecca L. 1984. "Creating Neighborhood Identity through Citizen Activism." *Urban Geography* 5: 49–70.

Smith, Rebecca L. 1985. "Activism and Social Status as Determinants of Neighborhood Identity." *The Professional Geographer* 37: 421–432.

Szelenyi, Ivan. 1983. *Urban Inequalities under State Socialism*. Oxford and New York: Oxford University Press.

Tilly, Charles, and Sidney Tarrow. 2006. *Contentious Politics*. Oxford, UK: Oxford University Press.

Tomba, Luigi. 2004. "Creating an Urban Middle Class: Social Engineering in Beijing." *The China Journal* 51: 1–26.

Tomba, Luigi. 2005. "Residential Space and Collective Interest Formation in Beijing's Housing Disputes." *The China Quarterly* 184: 934–951.

Vogel, Ezra F. 1990. *One Step Ahead in China: Guangdong under Reform*. Cambridge, MA: Harvard University Press.

Wang, Ya Ping. 2001. "Urban Housing Reform and Finance in China: A Case Study of Beijing." *Urban Affairs Review* 36: 620–645.

Wang, Ya Ping, and Alan Murie. 1996. "The Process of Commercialisation of Urban Housing in China." *Urban Studies* 33: 971–989.

Wang, Ya Ping, and Alan Murie. 2000. "Social and Spatial Implications of Housing Reform in China." *International Journal of Urban and Regional Research* 24: 397–417.

Wu, Fulong. 2001. "Housing Provision under Globalisation: A Case Study of Shanghai." *Environment and Planning A* 33: 1741–1764.

Wu, Fulong. 2002. "China's Changing Urban Governance in the Transition towards a More Market-oriented Economy." *Urban Studies* 39: 1071–1093.

Yip, Ngai-ming, and Yihong Jiang. 2011. "Homeowners United: The Attempt to Create Lateral Networks of Homeowners' Associations in Urban China." *Journal of Contemporary China* 20: 735–750.

Zhang, Li. 2008. *Private Homes, Distinct Lifestyles: Performing a New Middle Class*. Edited by Li Zhang and Aihwa Ong, *Privatizing China: Socialism from Afar*. Ithaca, NY: Cornell University.

Zhu, Yushu, Werner Breitung, and Si-ming Li. 2012. "The Changing Meaning of Neighbourhood Attachment in Chinese Commodity Housing Estates: Evidence from Guangzhou." *Urban Studies* 49: 2439–2457.

Appendix 1. Questions used to generate factor scores of neighborhood perception

Neighborhood perception	Questions used in the questionnaire
Perceived civic engagement	Would you say (it is very likely, likely, neither likely nor unlikely, unlikely, or very unlikely) that you would engage in the following neighborhood affairs: E1 If there is a problem affecting the residential environment of the whole community, would you *take the initiative* to solve the problem with other residents? E2 If someone else takes the initiative to solve the neighborhood problem, would you get engaged? E3 If a public project in this community does not directly involve your interests, would this project be worth your time? E4 If a public project in this community does not directly involve your interests, would this project be worth your money?
Informal social control	Would you say (it is very likely, likely, neither likely nor unlikely, unlikely, or very unlikely) that you would try to solve or intervene in the following scenarios occurred in this neighborhood: E5 Deliberately damaging public facilities, such as elevators or fire hydrants (vandalism) E6 Impairing the neighborhood environment, such as spitting or graffiti E7 Conflicts among neighborhood residents, such as quarrels and fights
Consciousness of property rights	Would you say (you are familiar with, somewhat familiar with, less familiar with/have heard of, have never heard of) the following aspects of neighborhood governance and related laws: D1 Maintenance fund D2 Collective revenues shared by homeowners D3 Partitioned ownership of condominium D4 Revenues and expenditure of property management fees D5 Expenditure of revenues from homeowners' common property (such as parking lots or outdoor advertisements) D6 Laws and regulations relevant to residential experience such as Real Right Law and Regulation on Realty Management

Homeowner associations and neighborhood governance in Guangzhou, China

Shenjing He

Department of Urban Planning and Design, The University of Hong Kong, Pokfulam, Hong Kong

Drawing on a large-scale household survey and in-depth interviews in Guangzhou, China, this research closely examines the formation and operation of homeowner associations (HOAs) and their governance efficacy in urban neighborhoods. This study examines the application of theories on private governance and collective action in the Chinese context and integrates the empirical analyses into a general discussion of state–market–society interactions. The HOA concept in China is far from a form of private governance by Western definition. On the contrary, it generates a societal force to counterbalance the market force brought by property management companies and the state power enforced by the residents' committee to address homeowners' political and material needs. Collective action theory is a useful explanatory tool for the HOA's governance efficacy in China. Yet, the state power and market force also bring a strong imprint on the decision-making of collective action in terms of determining organizational arrangements, the availability and quality of information, and the benefits of collective action.

1. Introduction

Common-interest developments (CIDs) providing homes with shared amenities were on the rise in post-World War II USA. These residential developments fed into the multiple needs of the fast-growing middle class seeking high-quality suburban homes, developers striving to build more housing units on less land, and municipalities aiming to increase the tax base while avoiding the cost of providing public goods and services. The emergence and prevalence of CIDs signified the rise of private neighborhoods and private governance. The term "privatopia" precisely describes this form of housing development and neighborhood governance that is dominated by private developers and attracts middle-class homebuyers with exclusive and tailored housing choices (McKenzie 1994). Within these CIDs, homeowner associations (HOAs) are set up by the developers. Homebuyers automatically become a member of the HOA and are bonded to certain deed restrictions, covenants, and conditions to govern their neighborhoods, including the use of property and other aspects of community life. A HOA in the USA is therefore described as "a private government" run by the homeowners and supported by attorneys and other professionals (1994). Along with the unprecedented development of private neighborhoods, HOAs (also known as neighborhood associations or neighborhood councils) have become increasingly widespread and emerged as a norm for governance in private neighborhoods in the USA[1]. The prevalence of HOAs reflects a further step of privatization and the retreat of the state in urban governance (McKenzie 2005). In the US context, the role of the HOA in neighborhood governance is mainly concerned with public participation/collective action and private governance (Cooper and Musso 1999; Dilger 1992; Glasze 2005).

Studies on HOAs have gone beyond the US context to examine this widespread phenomenon in Russia, India, Turkey, South America, and East Asia (Caldeira 2000; Chen and Webster 2005; Erman and Coskun-Yıldar 2007; Polishchuk and Borisova 2010; Yip and Forrest 2002; Yip and Jiang 2011; Zérah 2007). In the East Asian context, the HOA also induces heated debates on collective action and private governance. For instance, HOAs in Taiwan are set up by or under the support of developers and property management companies (PMC) and are deemed a more effective form of private governance, whose capacity outweighs that of either government (by coercion) or residents (by voluntary association) (Chen and Webster 2005, 2006). Evaluation of HOAs or owners' corporations in Hong Kong is less positive. The anti-democratic organizational arrangement of owners' corporations tends to limit, rather than enhance participation of ordinary owners and even exacerbates socio-spatial segregation in those typically large-scale, high rise, and technically complex environments (Yip and Forrest 2002). In mainland China, the HOA has brought about significant changes to neighborhood governance (Read 2003, 2008; Yip and Jiang 2011). Urban neighborhoods that used to be managed by either danwei (work unit) or juweihui (residents' committee [RC]) are now exposed to the complex power relations among real estate developer, PMC, homeowners, and HOAs. As important components of China's changing urban governance, the growing HOAs and the empowered homeowners have become a rising social power at neighborhood level. However, the development of the HOA in China is still highly contingent upon its interplay with the persisting state power and the burgeoning market force.

As a nascent phenomenon emerging amidst the housing reform, the HOA in China remains an under-researched topic. Yet, the Chinese case holds the potential for enriching our understanding of HOAs and neighborhood governance so far mainly deriving from the Western context, where the discussion on exogenous power – especially the state power – is insufficient since state interference is rarely an important concern except for preventing and monitoring misconduct and corruption (McKenzie 2005). Building on the extant literature, this study combines the analytical perspectives of private governance and collective action to scrutinize the role of HOAs in China's urban neighborhood governance. Drawing on a large-scale household survey and in-depth interviews in a number of gated urban neighborhoods in Guangzhou, this research examines the theories on private governance and collective action in the Chinese context and integrates the empirical analyses into a general discussion of state–market–society interactions. The rest of the paper is organized as follows: section two reviews major debates related to the development and governance of the HOA in private neighborhoods; section three gives a detailed account of China's emerging HOAs and neighborhood governance; section four presents an empirical study on the formation and operation of HOAs in Guangzhou and examines the governance efficacy of HOAs by taking account of the influences from the state, market, and society; and section five concludes.

2. Debates on the HOA and its role in neighborhood governance

As a non-government organization executing important duties of neighborhood governance, the HOA is considered a bottom-up urban governance mode and is therefore conducive in fostering civil engagement by mediating between individual home-owners/citizens, on the one hand, and large-scale bureaucracies (e.g. municipalities and other state agents) as well as developers and other business interests, on the other (Cooper and Musso 1999). Yet, its role in neighborhood governance remains controversial. While a HOA is considered to be a more democratic and effective form of small-scale governance efficiently providing collective goods/services and fostering public participation and community attachment (Cooper and Musso 1999; Dilger 1992; DiPasquale and Glaeser 1999; Rosenblum 1998), it is also widely criticized for its over-reliance on market forces that often result in residential segregation and social inequality (McKenzie 1994, 2005; Yip and Forrest 2002).

Studies on HOAs are closely related to a wide range of theoretical debates, such as civil society and grass-roots democracy, private neighborhood/gated communities and private governance, and collective action theory. This study mainly aims to understand the role of HOAs in neighborhood governance and therefore focuses on the latter two bodies of literature. The first strand of literature views the HOA as a form of private urban governance associated with the globally prevalent private neighborhoods and gated communities. Glasze (2005) applies club economics theory to explain the wide-spread private neighborhood at a global scale and its economic and political organization. The self-administration of private neighborhoods through HOAs is believed to be able to solve the free-rider problem for collective goods and render them excludable. It is therefore considered as an institutional innovation for efficient provision of local public goods (e.g. private golf courses) for a certain price to be paid by inhabitants (Beito, Gordon, and Tabarrok 2002; Foldvary 1994). These excludable collective goods provided within the private neighborhoods on the basis of ownership–membership are called "club goods" (Buchanan 1965). Buchanan's club theory demonstrates the possibility of efficient delivery of local public goods via clubs. Along the same vein, Glasze (2005, 224) interprets the establishment of private neighborhoods with their self-governing organization as "the creation of club economies with territorial boundaries." This also explains the prevalence of private neighborhoods and HOAs among developers, local governments, and inhabitants (Glasze 2003; McKenzie 1994). In many private neighborhoods, the HOA literally becomes the "private government," which possesses the legal right to keep homeowners' property under detention, to organize elections, to tax homeowners, and to execute penalties (McCabe 2011). To a large extent, the prevalence of HOAs in the US and other market economies is based on the presumption that the market can tackle the problems of public goods and externalities and therefore offers more flexibility and efficiency than municipal governments. This also reflects the demise of the omnipotent state and the rise of a minimal state resulting from deregulation and privatization that prevail globally (Glasze 2003, 2005). Some scholars see HOAs as a necessary and useful means to promote communal ownership and community self-management and therefore conducive to urban political economic and environmental development (e.g. Nelson 1999), while others contend that the efficiency of private neighborhoods and HOAs lies in integrating property owners and the provider of local collective goods (e.g. Deng 2002). In this regard, private neighborhoods and HOAs are considered to be the biggest experiment for self-administration, as they are based on incumbent residents' deep understanding of community and its resources, which endows HOAs a high degree of flexibility and resilience in neighborhood management (Nelson 2005).

The second strand of literature relates the formation and operation of HOAs to the theory of collective action. Known as the "zero contribution theory," Olson (1965) contends that self-interested rational individuals would not contribute to the production of public goods and "selective incentive" is necessary to induce collective action. Such selective incentive can either be in the form of one agent having a greater interest in the collective action or having a group of agents responsible for making others cooperate by reward or punishment (1965). Conventional collective action theorists point out that people's behavior in collective action is affected by many structural variables including size of group, heterogeneity of participants, their dependence on the benefits received, the type and predictability of transformation processes involved, the nesting of organizational levels, monitoring techniques, and the information available to participants. However, Ostrom (1998) suggests that in one-shot or finitely repeated dilemmas, structural variables do not affect levels of cooperation at all. As a supplement to the conventional collective action theories, personal attributes including reciprocity, reputation, and trust can help individuals to overcome the strong temptations of short-run self-interest so as to achieve results that are "better than rational" (Cosmides and Tooby 1994; Ostrom 1998).

The importance of authorizing citizens to constitute their own associations using their own knowledge and experience concerning the public problems they face has now been widely recognized. In this sense, the HOA creates a mechanism that allows individual resident's active participation in the design of institutions, affecting them to solve collective action problems themselves, although the institutional arrangement of the HOA might be manipulated by external forces such as developers and local government, in some cases. Meanwhile, a strong belief is held that the institutional structure of private neighborhoods and HOAs can reduce the transaction costs involved with undertaking collective action since the closely bonded social network among homeowners enables them to inform and motivate members of the community in an effective manner (Beito, Gordon, and Tabarrok 2002; Groves 2006).

These two strands of literature are of significance to understand not only the decision-making on the formation of HOAs, but also the internal functioning of HOAs and their role in neighborhood governance. For instance, the collective action paradox plays a role not only ex ante, but also ex post the formation of a HOA. A study of HOAs in Russia shows that the ability to make collective agreements is crucial for the establishment as well as the efficiency of a HOA (Polishchuk and Borisova 2010). Similarly, the formation decision of a HOA in mainland China, Hong Kong, and Taiwan is subjected to a number of collective action problems, which are contingent upon neighborhood size, building type, cultural acceptance of private responsibility in community management, strength of enabling legislation, strength of municipal government, and neighborhood homogeneity (Chen and Webster 2005; Lai and Chan 2004; Read 2008). Meanwhile, studies in the USA point out that private governance executed by HOAs is not always effective. The role of HOAs in neighborhood governance, particularly in fostering grass-roots governance and democracy, is largely constrained by its property ownership-based organization and is often manipulated by a powerful coalition of lawyers, property managers, accountants, and others (McKenzie 2005). Therefore, the quality and extent of HOA members' participation in private neighborhood governance are very limited (Low 2003; Rosenblum 1998). It is worth noting that although the theories of private governance and collective action mainly deal with economic and institutional issues concerning the decision-making and internal functioning of the HOA, they are also of significance to analyze the influence from the political arena, e.g. state interference.

To assess the governance efficacy of neighborhood-based initiatives, Chaskin and Garg (1997) map out several critical issues: the relationship between these initiatives and local government; issues of representation, legitimacy, and connection; and long-term viability. These issues are pertinent to the theories of private governance and collective action and have provided some guidelines to understand the role of the HOA in neighborhood governance. For instance, a recent study on the performance of HOAs in Russian cities applies the collective action theory to illustrate that the interplay between institutions, social capital, and organizational governance underpins the success of a HOA (Polishchuk and Borisova 2010). In general, governance efficacy can be assessed by the quality of service provided by the governing body and the effectiveness of its operation. Unlike corporate governance, neighborhood governance efficacy is difficult to be measured quantitatively. Criteria-based assessment and interpretive approach are frequently applied to assess democratic performance of governing organizations in the field of public administration (Mathur and Skelcher 2007), while in neighborhood governance, influence on fostering civic engagement and participatory governance (Cooper 2005; Hasson and Ley 1997) and/or residents' subjective evaluation/perception (Fu et al. 2015) are commonly employed to appraise governance efficacy.

3. HOA and neighborhood governance in the Chinese context

In the extant literature on HOAs and neighborhood governance, the role of the state is often overlooked since the HOA is generally considered as an emerging form of private governance resulting from the retreat of the local state in urban governance under the increasingly neoliberalized market economy. However, to understand the development of HOAs in Chinese cities, the state's indispensable role in neighborhood governance needs to be fully recognized.

In the wake of economic reforms, the urban governance system in China has seen the reinvention and consolidation of territorial organizations (Wu 2002). Specifically, a hierarchy of territorial organizations at the city level, that consisted of municipal government, district government, street offices (subdistrict offices), and RC, started to emerge and brought about a vertical integration of grass-roots state agencies and local governments (Bray 2006). In the pre-reform era when danwei was the basic unit governing community affairs, the RC was responsible for urban individuals without formal workplace affiliations. Since the launch of economic reforms, the power of the RC has been consolidated as a strategy to restore the Party's political authority over urban space. As noted by Davis (2006), the RC system embodies a nested honeycomb of party–state control extended into every urban neighborhood to fill the void in social control and social service provision in the post-danwei era. In other words, the explanation for the revitalization and consolidation of the RC system is twofold: it epitomizes the state's political desire to regain control at the grass-roots level and it brings a countervailing force against the retreat of residents from public life (Yip 2012).

Since the 1980s, commodity housing projects were started to be implemented in full swing to replace the former housing structure dominated by a danwei compound and traditional open access neighborhoods in urban China (Li 2003). Among these housing developments, the majority are gated and walled. Gated commodity housing estates gradually became a dominant residential form in urban China (He 2013). To a great extent, these housing estates are very similar to the common-interest developments in the USA. At present, 80–90 percent of newly constructed houses in Chinese cities are in gated communities. Along with the massive development of gated communities, the first PMC was established in 1981 in Shenzhen, the avant-garde city of China's market reforms. Ten years later, the first HOA in China also emerged in Shenzhen to solve the conflicts between homeowners and the PMC.

Resultantly, urban governance in China has seen the diversification of governing bodies, including market agents and bottom-up organizations, in addition to state agents at various levels. Appearing as the three pillars of neighborhood governance, RCs represent the state at the neighborhood level, PMCs represent the market, and HOAs represent the societal force. Yet, the power relations among the three actors are highly uneven and complicated. As an active grass-roots government agent extended from higher level state apparatus, the RC retains its dominant position in neighborhood governance, although its responsibilities and power have been partly taken over by other stakeholders. For instance, developers are required to provide office space for the RCs according to government regulations (Davis 2006). In Shanghai, state control is penetrating into upscale, gated communities through various strategies employed by the RC, including forming a local neighborhood governance coalition with the PMC, gaining support from the HOA, and hosting special events and activities targeting elderly residents (Sun and Yip 2014). In the meantime, the prevalence of gated communities has given rise to the PMC, which partly takes up the responsibility of the RC to provide collective goods and social services such as sanitation and public security. The PMC also provides tailored services such as landscaping and gardening, maintenance of facilities, and other club goods by charging monthly property management fees. In many cases, PMCs are affiliated to or hired by developers that maintain reciprocal relations with local governments, and therefore their decisions in property management affairs can be easily supported by government organizations, including the RCs (Fu and Lin 2014).

A homeowners association in China is different from its US counterpart in many ways, in particular in the lack of full legitimacy and the separation from and constant strife with PMC. Almost every state in the USA has its own legislation for HOAs, whereas in China, the first regulation relevant to HOA is the "Property Management Regulation" published in 2003 and revised in 2007 by the State Council, in which a HOA is loosely defined as a self-managed organization whose duty should not go beyond property management affairs. In 2009, the Ministry of Housing and Urban-rural Development issued a more detailed regulation entitled "Guideline for Homeowners' Assembly and HOA." This new regulation clarifies the relationship between HOAs and different government agents. The HOA is a self-organization parallel to the RC, both being supervised and monitored by the street office. Yet, a HOA should support the RC and accept its guidance and supervision (Clause 54). According to the regulation, the formation and operation of a HOA, especially at the preparatory stage, are subjected to close interference by the street office and RC, including direct participation in the preparatory group and nomination of homeowner delegates. The regulation details the procedures to form a HOA as well as its obligations and daily activities. However, its independent legal status has not been fully recognized. Apart from a vague statement, "the street office should process complains from HOA in time" (Clause 49), no specific legal and institutional support have been stipulated. More often than not, when a HOA seeks support from the street office concerning conflicts with the PMC, its complaints and charges are not always dealt with in a timely and even-handed manner, especially in the case where the PMC is a subsidiary of a powerful developer maintaining good relationship with the local government. In addition, different from other contexts such as in the USA and Taiwan, where the HOA is either integrated or closely coordinated with the PMC, China's HOAs are separated from PMCs and often emerge as a counter-force to the PMC. In many Chinese cities, conflicts and disputes between HOAs and PMCs are ubiquitous. Overall, RCs possess a strong position in neighborhood gover-nance (Bray 2006; Yip 2012). A PMC also has the upper hand in manipulating neigh-borhood management affairs, especially those upholding a close relationship with local government (Fu and Lin 2014). More often than not, a HOA is subordinate to the RC and the PMC or the alliance between them. However, the power relation among the three stakeholders is by no means static, and its impact on neighborhood governance remains unexplored.

HOAs in China come in various forms and with different characteristics. Read (2008) has summarized different forms of HOAs in China, which are determined first, by the degree of autonomy from and leverage over external actors (i.e. state and market agents); second, by internal practices and relationships with their constituents; and third, by the exercising of power, mainly through collective action. This is in accordance with the theories of private governance and collective action against the backdrop of complex state–market–society relations.

Subject to a number of factors, the development of HOAs in different Chinese cities varies considerably. For instance, over 80 percent of urban neighborhoods in Shanghai have set up HOAs, while according to the Guangzhou Land and Housing Management Bureau, only 25 percent of residential neighborhoods in Guangzhou had established HOAs by the end of 2013. The reasons are rather complex and very much depend on local institutional, social, and housing market conditions. In Shanghai, the development of HOAs is highly encouraged by the local housing authority and actively assisted by the street offices and RCs. Not surprisingly, those HOAs formed under state control tend to have a close relationship with the local government (Yip and Jiang 2011). In the Shanghai case, heavy government involvement has been employed as a major mecha-nism to overcome the collective action problem to establish HOAs. In contrast, in Guangzhou, it was reported that the establishment of HOAs in many neighborhoods had been held back by strong opposition from developers and PMCs, whereas institutional support was absent (Southern Metropolis Daily 2013). Lacking institutional and market inducements, development of HOAs in Guangzhou is therefore seriously deterred.

In spite of the dearth of literature directly dealing with HOAs in China, recent years have seen a proliferation of scholarly works on a highly relevant topic, that is, gated communities in China. Echoing international debates on gated communities and HOAs, relevant studies in China can be largely summarized into three strands of arguments. The first set of research works connects the formation and operation of HOAs with the rising political demands of homeowners that lead to neighborhood activism and different forms of grass-roots democracy (Davis 2006; Kelly 2006; Perry 2008; Read 2003, 2008; Tomba 2005; Yip and Jiang 2011). Restored private property rights and increased homeownership have fostered awareness of consumer rights and induced numerous con-sumer-/homeowner-related activisms (Davis 2006). The most common form of home-owner activism is for homeowners to defend their property or other material rights within or beyond their neighborhoods against the misconducts of more powerful market players (Read 2008; Yip 2012). However, it is not uncommon for owners' power to be circumscribed when they tried to advance their rights in opposition to the developer/ PMC or the local authority through the HOA (Davis 2006; Yip and Jiang 2011).

The second theme emphasizes the indispensable role of the state in facilitating the rapid development of gated communities and the emergence of HOAs (see Huang 2006; Yip 2012). The prevalence of gated communities is considered a result of the downscaling of urban governance and the enforcement of political control at the neighborhood level (Huang 2006). Thirdly, other commentators tend to see the rising market force as a useful explanatory tool for the prevalence of gated communities in China (see Webster, Wu, and Zhao 2006; Wu 2006). They emphasize the importance of market efficiency in the provision of exclusive housing development, club goods, and the organization of neighborhood management affairs. A gated community is therefore considered to form a club realm of consumption so as to efficiently provide high-quality housing and tailored services to the emerging affluent population (Wu 2006). Meanwhile, a HOA is believed to be able to offer more effective and pliable management than local government (Chen and Webster 2005).

Nevertheless, these studies remain insufficient in quantitatively demonstrating the internal functioning and governance efficacy of HOAs, and they are incomplete in the coverage of multiple politico-economic and social issues. In the Chinese context, where an unfledged market and the nascent civil society meet the less omnipotent yet still powerful state, the interplay among the three forces displays both features of path dependency and contingency. Therefore, a thorough understanding of the HOA in China calls for a meticulous examination of how the interrelationships among the state, market, and society affect the formation and operation of the HOA and its governance efficacy.

4. An empirical investigation in Guangzhou

Data used in this research are from a large-scale survey conducted in Guangzhou from 2011 to 2012. This is the first large-scale survey purposely examining the formation and operation of HOAs in China, which will enable a close scrutiny of the role of HOAs in neighborhood governance.

In this study, HOAs were selected using a multistage stratified random sampling method. In the first stage, three primary sampling unit strata in Guangzhou were determined by land use and residential density: inner core, inner suburb, and outer suburb. In the second stage, street offices within each stratum were selected with reference to the total number of street offices located in each stratum and the spatial distribution of these street offices in Guangzhou. In the third stage, based on a list of neighborhoods with an operating HOA provided by the South China Centre for Harmonious Community Development, one neighborhood and a list of 2–8 adjacent neighborhoods (the exact number depending on the size of these neighborhoods) were chosen within a selected street office area. When interviewers failed to get access to the targeted neighborhoods due to difficulties such as gatedness, intervention from PMCs, or frequent failures in recruiting a reasonable number of cooperative residents, an alternative neighborhood would be selected from the list of adjacent neighborhoods. Because residential neighborhoods located in the outer suburb area rarely establish HOAs, the outer suburb area has fewer number of neighborhoods included in this study.

In total, 39 neighborhoods were selected for the survey. All of them are gated and guarded. Figure 1 shows the distribution of surveyed neighborhoods. In the fifth stage, residents within a neighborhood were recruited based on their home address using a fixed interval. The number of surveyed residents within a given neighborhood was adjusted according to the neighborhood size. His or her neighbors were chosen as a respondent if a resident refused to participate in this survey. Finally, 1809 valid

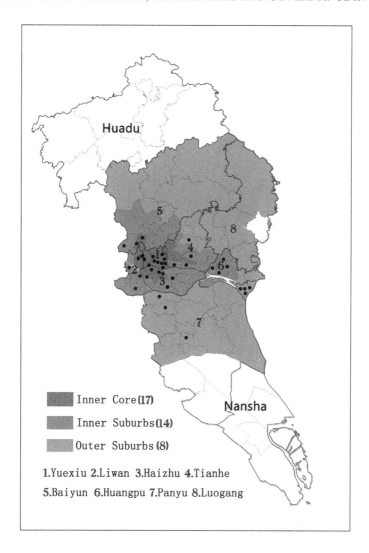

Figure 1.   Distribution of surveyed neighborhoods.
Note: Figures in the parentheses refer to the number of surveyed neighborhoods in each area.

questionnaires were yielded. In addition, in-depth interviews were also conducted with directors and board members from 69 HOAs within gated communities, which include 31 in the inner core, 20 within the inner suburb, and 18 in the outer suburb. The selection of these 69 HOAs, which is inclusive of the 39 surveyed neighborhoods, also followed multistage stratified random sampling method.

4.1.   Formation and operation of HOAs in Guangzhou

According to our interviews, among the 69 neighborhoods covered by this study, 71 percent set up HOAs after 2007. This is, to a great extent, related to the publication of the Property Rights Law and the revised Property Management Regulations. This suggests that legislative and institutional support is of great significance to the development

Table 1. Profile of interviewed HOA directors/board members.

	No.	Percent
Educational attainment		
Higher education (college and above)	45	65
Middle school	21	31
Primary school and below	3	4
Average housing floor area (m²)	103.1 (household)	
Party membership	No.	Percent
Yes	19	49
No	20	51
Employment status	No.	Percent
Retired	40	58
Employed	29	42
Occupation (current or latest)	No.	Percent
Administration	19	51.4
Professional	11	29.7
Clerical staff	0	0
Service, manual labor, etc.	7	18.9
Housing ownership	98 percent	

	No.	Percent
Gender		
Female	8	21
Male	31	79
Average age		57.5
Age	No.	Percent
Under 40	2	5
40–50	10	26
50–60	13	33
Above 60	14	36
Annual income per capita	No.	Percent
No answer	4	5.8
Under 30K	14	20.3
30K–50K	19	27.5
Above 50K	32	46.4
Hukou	No.	Percent
Local	60	87
Non-local	9	13

Table 2. Funding source of HOAs.

	Percent	Monthly amount (RMB)	Funding sources (cases)							
			Property management fee	Communal income	Self-raised fund from homeowners	Donation from homeowners	Provided by PMC	Provided by developers	Others
Funds for daily operation	62.3	10,634	10	22	2	4	5	0	2
Subsidies for HOA board members	20.9	2309.2	5	6	0	0	3	0	0
Salary for HOA board members	9	3016.6	4	4	2	2	2	0	0

of HOAs, even though the legal status of HOAs remains rather fuzzy. Table 1 shows the basic information of 69 interviewed HOA directors and board members. About 80 percent of the board members are male and 65 percent of them have gone through higher education. This is a comparatively aged group, with an average age of 57.5 and about 60 percent of retirees. The majority of them belong to managerial staff or professionals. Their average annual income per capita is relatively high, with 78 percent reporting a figure above RMB 30,000, which is well beyond the average level of RMB 25,000 per capita in Guangzhou in 2013. In addition, they also enjoy better housing conditions. The majority of them hold a local Guangzhou hukou, and about half of them have party membership. In general, these HOA directors and board members belong to an elderly elite group and have comparatively high socioeconomic and political status and are able to devote sufficient time and effort to the voluntary work for the HOA.

As stipulated in the 2009 regulation, throughout the preparation and establishment of a HOA, the street office and the RC have played an important role in supervising and nominating candidates for the HOA preparatory group members and HOA board members. Financing is a critical issue for the daily operation of HOAs. According to the "Guangdong Province Property Management Regulation," the costs of HOA's daily operation should be borne by all homeowners, while the forms of fund-raising and fund management will be decided by the homeowner assembly, and the HOA should publicize the usage of their operating fund on a seasonal basis.

As a matter of fact, many HOAs have difficulties to make ends meet. Table 2 shows the major funding source of 69 HOAs in Guangzhou. While 62.3 percent received regular funds for daily operation, only 20.9 percent managed to provide subsidies for HOA board members. Even fewer managed to pay salary to their board members. In addition, among the 69 HOAs, only 17 percent can afford a full-time assistant, while 42 percent do not even have their own office space.

In most cases, funding for HOAs comes from communal income of the neighborhood, such as rental income of business premises. For some HOAs, funding also comes from property management fees, especially for those ones without a PMC. Other possible funding sources include a maintenance fund, self-raised funds from homeowners, and donations from homeowners. In some rare cases, the daily operation of a HOA is supported by a PMC. Notably, our interviewees reported that neighborhood's communal income and maintenance fund are controlled by the PMCs in most cases. HOAs have to negotiate with PMCs to gain access to these funds. More often than not, this has become a major source of disputes and confrontations between HOAs and PMCs. Overall, the lack of stable funding sources has become a common predicament for the establishment and operation of HOAs in China. Throughout the processes of establishing and managing the HOA, board members have to invest a huge amount of time and effort. In the meantime, they have to face possible conflicts with PMCs and misunderstandings from homeowners. As collective action theorists point out, without certain selective incentives, especially necessary financial support, the daily operation of a HOA and its governance efficacy would be seriously jeopardized.

According to our interviews, the first task launched by HOAs after establishment is usually related to PMCs, either renewing a contract with the original PMC (26.9 percent) or firing the old PMC (29.9 percent). In the latter situation, a HOA would hire a new PMC in most cases, whereas some neighborhoods would opt for self-management. However, firing a PMC is not always an easy task. It was reported by our interviewees that in many cases, PMCs refused to move out the neighborhood and colluded with developers and local government to oppose the newly established HOAs.

Table 3. Power relation between HOA and other stakeholders.

	Relationship					Power relation						
	Very good	Good	Fair/ normal	Poor	Very poor	Much more powerful	Slightly more powerful	Much less powerful	Slightly less powerful	Equivalent	Contractual	Null
Residents committee	10.1	39.1	44.9	5.8	0	8.7	40.6	2.9	4.3	34.8	0	8.7
Developer	3.0	4.5	43.9	21.2	24.2	4.3	0	2.9	5.8	10.1	26.1	50.7
PMC	4.4	23.5	48.5	14.7	8.8	2.9	2.9	7.2	18.8	29.0	30.4	8.7
Other government agent	5.9	22.1	67.6	2.9	0	20.3	36.2	4.3	4.3	20.3	0	13.0
Homeowners	2.9	66.2	30.9	0	0	21.7	4.3	5.8	10.1	55.1	1.4	0

Other tasks tackled by HOAs right after establishment include solving problems related to communal properties, such as parking lots and commercial services (17.9 percent); solving problems related to public services, such as water and electricity supply (7.5 percent); and amending rules and code of conduct for neighborhood management. As for daily operation of HOAs, 71.2 percent of interviewees reported that their HOA has regular meetings. The average frequency of HOA board meetings is once every 5.1 weeks, and the majority of board members attend these meetings. Frequent meetings with non-board members were reported by 42 percent of interviewees and 48.5 percent believe that more than half of the homeowners in their neighborhoods care about their HOA. These figures indicate that most HOAs function well, and they actively interact with homeowners.

When interviewees were asked to evaluate the relationship between their HOA and other organizations, an antagonistic relationship between HOA and developer or PMC can be observed (see Table 3). Throughout the formation and operation of HOAs, it was reported that various interferences from PMCs and developers are very common. Under such circumstances, many homeowners are discouraged from participating in activities organized by the HOA, including voting within the homeowner assembly. Even if HOAs appeal to the street office, their requests are rarely properly attended or responded to promptly. On the other hand, HOAs have tried to maintain a good relationship with homeowners and government agents. When HOA directors/board members are asked to evaluate the power relations between the HOA and other stakeholders, street office and other government agents are considered to be more authoritative or possess more power over the HOA, whereas a PMC is considered less powerful, or at best equivalent to the HOA. However, some interviewees consider that the HOA and PMC/developer have a purely contractual relationship. Notably, a HOA is far from being superior to the PMC, in most cases, since the latter possesses better economic power and often colludes with the developer and local government. The fact that the PMC is the de facto and legal managing unit of urban neighborhood largely impairs the power of the HOA, especially when their goals are in conflict. Nonetheless, these evaluations suggest HOA's obedience to government agents and a clear competitive relation with PMCs. Homeowners are generally perceived as an equal party in relation to HOAs, and, in some cases, are considered to possess more power. The power relations among different stakeholders are also reflected in the number of major disputes that occurred in the neighborhood within the most recent five-year period. According to the interviewees, the numbers of property management-related disputes (69 cases) and developer-related disputes (68 cases) are the largest, followed by disputes among homeowners (60 cases). In comparison, the numbers of disputes related to local government (29 cases), HOAs (14 cases), and cross-neighborhood activism (3 cases) are much smaller. These findings suggest that a HOA intentionally maintains a good relationship with government agents, while competition and hostility between HOAs and PMCs are evident.

In summary, HOAs in Guangzhou remain as bottom-up, self-funded, well-functioning, and active self-organizations. It is therefore appropriate to contend that the HOA has started to emerge as an important societal force in neighborhood governance. Yet, the formation of HOAs is still strongly interfered with by the street office and the RC, and the operation of HOAs generally suffers from the lack of legislative protection and financial support. Meanwhile, the confrontation between HOA and PMC/developer marks the major difference to its counterparts in other contexts, where PMC and HOA are either incorporated or collaborative to generate a form of private governance. For instance, in Taiwan, motivated by the benefits of enlarging the market, seeking out

Table 4.   Principal component analysis of three major indicators of governance efficacy ($N = 1385$).

	Eigenvalue	Total variance explained	Factor loadings
Governance efficacy	2.042	68.056 percent	
• Accountability[a]: To what extent do you trust the HOA in your neighborhood?			0.795
• Representation[a]: To what degree do you think the HOA can represent homeowners' interests?			0.805
• Satisfaction[a]: To what extent are you satisfied with the HOA?			0.873

[a]These three indicators are measured by a single scale from 0 (lowest degree) to 5 (highest degree).

niches, and building favorable client–contractor relationships, PMCs have played an important part in helping neighborhoods establish HOAs, which can be deemed as a market response to the collective action problem deterring HOA formation (Chen and Webster 2005, 2006). In mainland China, such a market mechanism is still lacking. Overall, the formation and operation of a HOA are to a large extent determined by the uneven state–society and market–society relations. The lack of institutional support and market inducement has directly led to the underdevelopment of HOAs in China.

4.2.   Governance efficacy of HOA

Empirically, neighborhood governance efficacy can be assessed by its external influence on civic engagement and participatory governance (Cooper 2005; Hasson and Ley 1997) or/and internal functioning measured by residents' subjective evaluation (Fu et al. 2015). Governance efficacy of HOAs has rarely been quantitatively measured in existing studies, owing to the unavailability of empirical data and their difficulty to be quantified and standardized. In this study, governance efficacy of a HOA is appraised by the effectiveness of its internal operation measured by residents' assessment.

Referring to extant studies on assessing the performance and effectiveness of neighborhood governance (e.g. Chaskin and Garg 1997; Fu et al. 2015; Polishchuk and Borisova 2010), the value of HOA governance efficacy is obtained using principal component analysis of three indicators: accountability, representation, and satisfaction. Results from principal component analysis are shown in Table 4. All three indicators have high factor loadings. The extracted principal component (governance efficacy) explains 68 percent of total variance of the three indicators. These indicators are by no means exhaustive. Nonetheless, they have captured the most important outcomes of neighborhood governance to which the HOA is expected to contribute as far as the effectiveness of internal operation is concerned (Chaskin and Garg 1997). Similar questions are also asked to evaluate the performance of other organizations, including PMCs and RCs. Therefore, these indicators can provide a comparatively accurate measurement of the governance efficacy of HOAs by taking the influence of other governing bodies into account.

Table 5 outlines the profiles of respondents in the 39 neighborhoods. Among the 1385 respondents,[2] both male and female are well-represented. Their percentage of

Table 5. Profile of surveyed residents.

Category	No.	Percent	Category	No.	Percent
Educational attainment	No.	Percent	*Gender*	No.	Percent
Higher education (college and above)	727	52.5	Female	626	45.2
Middle school	542	39.2	Male	744	53.7
Primary school and below	115	8.3	*Average age*	46	
Average housing floor area (m²)	94.1 (household)	36.6 (per capita)	*Age*	No.	Percent
			Under 40	553	40.1
Party membership	No.	Percent	40–50	298	21.6
Yes	341	24.7	50–60	223	16.2
No	1041	75.3	Above 60	305	22.1
Employment status	No.	Percent	*Annual income per capita*	No.	Percent
Retired	415	54.6	No answer	74	5.3
Employed	756	45.4	Under 30K	489	35.3
Occupation (current or latest)	No.	Percent	30K–50K	215	15.5
Administration	613	47.8	Above 50K	606	43.8
Professional	296	23.1	*Hukou*	No.	Percent
Clerical staff	22	1.7	Local	1108	80
Service, manual labor, etc.	352	27.4	Non-local	277	20
Housing ownership	75 percent				

Table 6. Multilevel linear regression on governance efficacy of HOA (N = 1385).

	Model 1 (individual)	Model 2 (neighborhood characteristics)	Model 3 (neighborhood disputes)	Model 4 (HOA characteristics)	Model 5 (HOA's relationship with other agents)	Model 6 (HOA's power relation with other agents)
Individual attributes						
Gender	−0.049	−0.062	−0.075	−0.025	−0.034	−0.068
Age	5.500E-05	4.100E-05	3.800-E05	5.100E-05	2.200E-05	3.900E-05
Hukou	0.057	0.092	−0.021	0.050	0.023	0.058
Educational attainment	−0.048*	−0.040	−0.045	−0.059**	−0.047	−0.055*
Annual income per capita	8.23E-04	−7.230E-04	−7.710E-04	−8.670E-04	−1.228E-03**	−8.120
Party membership	0.122	0.190	0.180	0.199	0.095	0.164
Commercial housing ownership	−0.435***	−0.427*	−0.443*	−0.451**	−0.420**	−0.427*
Non-commercial housing ownership	−0.386	−0.349	−0.378	−0.479*	−0.310	−0.377
Residential period	−0.001	−0.001	0.001	−0.002	−0.012	−0.002
Neighborhood attachment	0.271**	0.224*	0.328**	0.252**	0.259*	0.274**
Neighborly relation	0.442***	0.450**	0.404**	0.414***	0.495***	0.423***
Overall built environment	0.414***	0.395**	0.386*	0.406**	0.288**	0.429**
Transparency of HOA operation	0.241**	0.268**	0.212**	0.225**	0.260**	0.255**
Participation in activities organized by RC	−0.129*	−0.119*	−0.132*	−0.134*	−0.207**	−0.120
Participation in activities organized by HOA	0.291***	0.276**	0.313***	0.303**	0.354***	0.278***
Participation in activities organized by PMC	−0.057	−0.054	−0.046	−0.068	−0.032	−0.049
Participation in activities self-organized by homeowners	−0.009	−0.005	0.027	−0.002	−0.030	−0.007
Participation in neighborhood politics	0.035**	0.035*	0.030*	0.035*	0.028	0.036**

Understanding of rules and laws on property management	−0.033	−0.050	−0.028	0.0136	−0.028	−0.040
Neighborhood characteristics						
Neighborhood size (population)		−1.220E-04**				
Degree of gatedness		−0.012				
Average housing price		0.009				
Age of neighborhood		−0.021				
Neighborhood disputes						
Developer-related disputes			0.051			
Property management-related disputes			0.021			
Government-related disputes			−0.053*			
Cross-neighborhood activism			−0.089			
HOA characteristics						
Voting rate in founding HOA				−0.007		
Number of HOA board members				0.072**		
Time to found HOA				−0.025**		
Frequency of HOA board meeting				0.004		
HOA's Relationship with other agents[a]						
RC and street office					−0.119	
Other government agents					0.183	
Developer					0.032	
PMC					−0.253**	
HOA's Power relation with other agents[b]						

(Continued)

Table 6. (Continued).

	Model 1 (individual)	Model 2 (neighborhood characteristics)	Model 3 (neighborhood disputes)	Model 4 (HOA characteristics)	Model 5 (HOA's relationship with other agents)	Model 6 (HOA's power relation with other agents)
HOA's power over RC and street office						−0.350*
HOA's power over other government agents						−0.166**
HOA's power over developer						0.226*
HOA's power over PMC						−0.120
Constant	0.139	0.554**	0.145	0.340	0.696	0.244*
R^2 (level-1)	0.354	0.340	0.343	0.337	0.360	0.354
R^2 (level-2)	0.210	0.238	0.140	0.397	0.340	0.300

Notes: *$p \leq 0.1$; **$0.1 < p \leq 0.05$; ***$0.5 < p \leq 0.01$.
[a]Relationship between HOA and other agents; scores from 1 to 5, with higher score indicating better relationship.
[b]Directors of HOAs were asked to assess the power relation between HOA and other agent: −2 means HOA is much less powerful than the other agent; −1 means HOA is slightly less powerful; 0 means an equal relation; 1 suggest HOA is slightly more powerful; and 2 means HOA is much more powerful.

higher educational attainment, party membership, annual income per capita, and housing floor area are lower than those of directors/board members, yet higher than the average level in Guangzhou. In terms of occupation, more than 70 percent of them are either managerial staff or professionals. In addition, 75 percent are homeowners. This is explained by the fact that gated communities generally accommodate a more affluent population compared with other types of neighborhoods such as danwei compounds, old urban neighborhoods, and urban villages.

Table 6 shows results from hierarchical linear regression models on HOA governance efficacy, which include independent variables at both individual and neighborhood levels. Model 1 includes only individual-level variables. Models 2–6 are the results of adding different types of neighborhood-level variables, including neighborhood characteristics, neighborhood disputes, the relationship between a HOA and other agents, and power relations between a HOA and other agents, respectively. These models aim to reveal the complex mechanism of neighborhood governance by taking endogenous factors (individual characteristics, neighborhood characteristics, and HOA characteristics), exogenous factors (external actors-related disputes, relationship, and power structure between HOA and other agents), and their interactions (through hierarchical linear regression) into account.

At the individual level, educational attainment and commercial housing ownership are significantly and negatively associated with governance efficacy, which suggests that better-off homeowners with higher education tend to be more critical of a HOA's performance. Meanwhile, neighborhood attachment and neighborly relations show statistical significance in a positive way because both variables are contributive to social cohesion and collective action. Overall assessment of built environment and transparency of HOA operation also significantly improve the perception of HOA performance.

Participation in activities organized by different organizations seems to have different influences on the evaluation of HOAs. Those residents frequently participating in HOA-organized activities tend to evaluate positively, while those participating in RC-organized activities tend to evaluate HOAs negatively. This suggests a competitive relation between HOA and RC in neighborhood governance. Additionally, those with strong political interests/appeals (i.e. frequently participating in neighborhood politics) tend to evaluate the performance of HOAs more positively.

In Model 2, educational attainment becomes insignificant after the introduction of neighborhood characteristics variables. Not surprisingly, larger neighborhood size suggests lower governance efficacy since a smaller community tends to foster closer social relation among its members and is therefore conducive for collective action (Miller 1992). In Model 3, again, educational attainment does not show statistical significance, while government-related disputes are significantly and negatively related to governance efficacy, which suggests that conflicts with local government would undermine the governance efficacy of a HOA. In Model 4, the number of HOA board members is significantly and positively related to governance efficacy, while time spent on founding the HOA shows negative connection, indicating that the efficiency of founding HOA significantly affects residents' evaluation of governance efficacy. In Model 5, interestingly, a HOA's relationship with the PMC is negatively associated with governance efficacy. On the one hand, this suggests that the confrontation between HOA and PMC could serve as a catalyst to foster community cohesion among homeowners so as to make possible collective action and enhance HOA's governance efficacy. On the other hand, this also suggests that the unsatisfactory club goods provided by the PMC offer opportunities for the HOA to defend its rights/interests on homeowners' behalf or to provide alternative collective services. In Model 6, a HOA's power relations with government agents and market agents demonstrate very different effects on its governance efficacy. Authoritative street offices and RCs help to promote HOA governance efficacy. In contrast, a less powerful developer is positively associated with governance efficacy. Although the interrelationship and power structure between HOAs and external actors (e.g. PMCs and government agents) unveiled by the regression models might not be completely accurate, owing to the nature of subjective evaluation provided by the respondents and the limited choice of relevant variables, these results do convey some important messages in regard to the uneven power relation between the state, market, and society. This unequal power structure is also supported by the interviews with HOA directors and board members, who reported several cases of collusion between the PMC and RC to squelch rights-defending activism organized by the HOA.

To summarize, the findings are threefold. First of all, although China's gated communities resemble those in the West in many aspects, the HOA in China does not really conform to the corresponding private governance mode. Instead of joining forces with the PMC and developer to provide tailored club goods for middle-class homeowners, a HOA generates a societal force to counterbalance the market force brought by PMCs and the state power enforced by the RC to address homeowners' political and material needs. This is most evident in the rivalry between the HOA and PMC/developer. In this sense, HOAs in China hold some potential for fostering civil engagement and grassroots democracy.

Second, collective action theory remains as a useful explanatory tool for governance efficacy of HOAs in China. The effectiveness of collective action is highly contingent upon (1) the size of the neighborhood; (2) group members' dependence on the benefits received, including their physical demands (e.g. overall quality of the built environment) and political demands (e.g. participation in HOA activities, neighborhood politics, and shared hostility against PMCs); (3) organizational arrangements (i.e. number of HOA board members and efficiency of founding the HOA; and (4) information available to participants (i.e. transparency of HOA operation). As contended by Ostrom (1998), besides these structural variables, personal attributes, in this case especially "trust," manifested by neighborhood attachment, neighborly relations, and social cohesion, also contribute to overcome collective action problems.

Thirdly, government agents have played a significant role in determining HOA's governance efficacy. For instance, conflicts and disputes with local government significantly impair the governance efficacy of a HOA. On the contrary, powerful street offices and RCs provide strong support for HOA governance. Meanwhile, a competition between state agents represented by the RC and societal forces represented by the HOA is tangible. This finding suggests a transforming neighborhood governance mode: HOAs will take up some governing functions that used to be dominated by the RC to play an increasingly important role in neighborhood management. Within this transforming process, rivalry between the two stakeholders is inevitable.

5. Conclusions

Making use of a large-scale household survey and in-depth interviews in neighborhoods with operating HOAs in Guangzhou, this study gives a detailed account of the formation and operation of HOAs and examines their governance efficacy. In order to provide a comprehensive understanding of the role of HOAs in China's neighborhood governance, this research examines the application of theories on private governance and collective action in China and situates the study within the context of state, market, and society interactions. Owing to the limitation of available empirical materials, the examination of exogenous factors, especially the state agents in this study, might not be sufficient. Yet, this study has tried to explore the interventionist role of the local state through examining the endogenous mechanism of the HOA. Indeed, the provision of club goods and the operation of collective action are by no means free of state intervention in the Chinese context. This is also one of the salient features of the Chinese urban society, where changes happening in the economic arena are always mingled with political interference.

In the wake of economic reforms, homeowners in China are emerging as consumer citizens who demand better housing quality and better collective goods and services. The emergence of HOAs feeds into the needs of homeowners to gain better control over their community lives, especially when the PMC fails to provide collective services reasonably or even harms their interests. It is worth noting that the HOA in China is far from a form of private governance by Western definition. In the Chinese case, a HOA is physically an integrated part of the gated community (private neighborhood) and is formed by homeowners. However, organizationally, it is separated from the private governance executed by the PMC. An antagonistic relationship between HOAs and developers/PMCs is commonly observed. In many cases, the confrontation between homeowners and PMCs directly triggered the establishment of HOAs. To a great extent, the HOA in China represents a grass-roots organization to countervail the imbalanced power dominated by the PMC in neighborhood management affairs. This entails a multifaceted impact of market force on HOAs. On the one hand, the rising market force enhances the awareness of homeowner's rights and leads to the establishment of HOAs; on the other, the insufficient or ineffective provision of club goods under the unfledged market conditions also instigates the need to set up a HOA. In short, the confrontational market–society relationship and the unsatisfactory market operation largely explain the different versions of private governance in Chinese cities.

It is undeniable that the emergence of HOAs signifies a growing social force in China's urban neighborhood governance. Empirical evidence has shown that participation in collective actions, especially confronting the PMC/developer and participating in neighborhood politics, in many cases, has become a direct impetus for the establishment of HOAs and usually leads to higher governance efficacy. The opportunities and constraints for HOAs to perform an active role in neighborhood governance can be largely explained by the collective action theory. As collective theorists contend, the formation and performance of a HOA are decided by a number of structural factors including neighborhood size, benefits generated by collective action, organizational arrangements, and availability and quality of information as well as some personal attributes, particularly trust and cohesion in this case. As an important self-organization formed by homeowners with raising consciousness of their property rights, HOAs bring a new opportunity for neighborhood self-management and grass-roots governance. It could be of great significance for the development of civil society in urban China.

Although this study mainly focuses on the internal operation of the HOA, empirical evidence suggests that the external influence of HOAs (e.g. nurturing civil society) remains unsatisfactory. As this study shows, much of HOA daily operations and obligations are still largely confined to property management affairs. The reasons are twofold. First, the lack of legal status and institutional and financial support has seriously hindered the HOAs from reaching out for broader civic affairs. Second, HOAs have to devote much of their efforts to confront PMCs and developers. In this regard, although the collective action theory is mainly concerned with endogenous factors, in the Chinese context, some exogenous factors (i.e. the state power and market force) also contribute a strong imprint on the decision-making of collective action in terms of determining organizational arrangements, the availability and quality of information, and the benefits of collective action.

Against the backdrop of administrative decentralization, urban governance in China has undergone fundamental transformation, which is best manifested in diversified governing agents and dwindling direct state intervention. These changes have given rise to HOAs in Chinese cities. Nonetheless, the development of HOAs is highly uneven, both across cities and within cities, since the establishment and performance of a HOA are highly contingent upon institutional arrangements, especially the legislative system and state involvement. Not only is the formation of HOAs closely monitored by state agents, but also their governance efficacy is significantly affected by the disputes and power relations between HOAs and state agents as well as the involvement of state agents in neighborhood activities. These findings suggest a persisting state influence throughout the development of China's HOAs.

Acknowledgment

I am grateful for the valuable comments made by professor Kam Wing Chan, professor Si-ming Li, and anonymous reviewers. I would like to thank Mr. Kun Wang and Miss. Jiaojiao Zuo for their assistance in data collection and data analyses.

Disclosure statement

No potential conflict of interest was reported by the author.

Funding

This work was supported by the National Natural Science Foundation of China [Ref. 41322003, 41271180].

Notes

1. According to the statistics from Community Associations Institute in the USA, in 2013, more than 26 million American households, involving approximately 66 million people, are living in neighborhoods governed by HOAs (source: http://www.caionline.org/info/research/Pages/default.aspx).
2. After removing questionnaires with missing values in the evaluation of governance efficacy, 1385 valid cases are included in this analysis.

References

Beito, David T., Peter Gordon, and Alexander Tabarrok. 2002. *The Voluntary City: Choice, Community, and Civil Society*. Ann Arbor: University of Michigan Press.
Bray, David. 2006. "Building 'Community': New Strategies of Governance in Urban China." *Economy and Society* 35: 530–549.
Buchanan, James M. 1965. "An Economic Theory of Clubs." *Economica* 29: 371–384.
Caldeira, Teresa P. R. 2000. *City of Walls: Crime, Segregation, and Citizenship in Sao Paulo*. Berkeley: University of California Press.
Chaskin, Robert J., and Sunil Garg. 1997. "The Issue of Governance in Neighborhood-based Initiatives." *Urban Affairs Review* 32: 631–661.

Chen, Chien-Yuan, and Chris Webster. 2005. "Homeowners Associations, Collective Action and the Costs of Private Governance." *Housing Studies* 20: 205–220.

Chen, Chien-Yuan, and Chris Webster. 2006. "Privatising the Governance and Management of Existing Urban Neighbourhoods." *Property Management* 24: 98–115.

Cooper, Terry L. 2005. "Civic Engagement in the Twenty-first Century: Toward a Scholarly and Practical Agenda." *Public Administration Review* 65: 534–535.

Cooper, Terry L., and Juliet A. Musso. 1999. "The Potential for Neighborhood Council Involvement in American Metropolitan Governance." *International Journal of Organizational Theory and Behavior* 1: 199–232.

Cosmides, Leda, and John Tooby. 1994. "Better than Rational: Evolutionary Psychology and the Invisible Hand." *American Economic Review* 84: 327–332.

Davis, Deborah. 2006. "Urban Chinese Homeowners as Consumer-Citizens." In *The Ambivalent Consumer*, edited by S. Garon and P. Maclachlan, 281–299. Ithaca, NY: Cornell University Press.

Deng, Feng. 2002. "Ground Lease-based Land Use System versus Common Interest Development." *Land Economics* 78: 190–206.

Dilger, Robert Jay. 1992. *Neighborhood Politics: Residential Community Associations in American Governance*. New York: New York University Press.

DiPasquale, Denise, and Edward L. Glaeser. 1999. "Incentives and Social Capital: Are Homeowners Better Citizens?" *Journal of Urban Economics* 45: 354–384.

Erman, Tahire, and Meliha Coskun-Yıldar. 2007. "Emergent Local Initiative and the City: The Case of Neighbourhood Associations of the Better-off Classes in Post-1990 Urban Turkey." *Urban Studies* 44: 2547–2566.

Foldvary, Fred. 1994. *Public Goods and Private Communities – The Market Provision of Social Services*. Aldershot: Edward Elgar.

Fu, Qiang, and Nan Lin. 2014. "The Weaknesses of Civic Territorial Organizations: Civic Engagement and Homeowners Associations in Urban China." *International Journal of Urban and Regional Research* 38: 2309–2327.

Fu, Qiang, Shenjing He, Yushu Zhu, Si-ming Li, Yanling He, Huoning Zhou, and Nan Lin. 2015. "Toward a Relational Account of Neighborhood Governance: Territory-based Networks and Residential Outcomes in Urban China." *American Behavioral Scientist* 59: 992–1006.

Glasze, Georg. 2003. "Segmented Governance Patterns – Fragmented Urbanism: The Development of Gated Housing Estates in Lebanon." *Arab World Geographer* 6: 79–100.

Glasze, Georg. 2005. "Some Reflections on the Economic and Political Organisation of Private Neighbourhoods." *Housing Studies* 20: 221–233.

Groves, Jeremy R. 2006. "All Together Now? An Empirical Study of the Voting Behaviors of Homeowner Association Members in St. Louis County." *Review of Policy Research* 23: 1199–1218.

Hasson, Shlomo, and David Ley. 1997. "Neighborhood Organizations, the Welfare State, and Citizenship Rights." *Urban Affairs Review* 33: 28–58.

He, Shenjing. 2013. "Evolving Enclave Urbanism in China and Its Socio-spatial Implications, the Case of Guangzhou." *Social & Cultural Geography* 14: 243–275.

Huang, Youqin. 2006. "Collectivism, Political Control, and Gating in Chinese Cities." *Urban Geography* 27: 507–525.

Kelly, David. 2006. "Public Intellectuals and Citizen Movements in the Hu-Wen Era." *Pacific Affairs* 79: 183–204.

Lai, Lawrence Wai-Chung, and Pearl Yik-Long Chan. 2004. "The Formation of Owners' Corporations in Hong Kong's Private Housing Estates: A Probity Analysis of Mancur Olson's Group Theory." *Property Management* 22: 55–68.

Li, Si-ming. 2003. "Housing Tenure and Residential Mobility in Urban China: A Study of Commodity Housing Development in Beijing and Guangzhou." *Urban Affairs Review* 38: 510–534.

Low, Setha M. 2003. *Behind the Gates: Life, Security, and the Pursuit of Happiness in Fortress America*. New York: Routledge.

Mathur, Navdeep, and Chris Skelcher. 2007. "Evaluating Democratic Performance: Methodologies for Assessing the Relationship between Network Governance and Citizens." *Public Administration Review* 67: 228–237.

McCabe, Barbara C. 2011. "Homeowners Associations as Private Governments: What We Know, What We Don't Know, and Why It Matters." *Public Administration Review* 71: 535–542.

McKenzie, Evan. 1994. *Privatopia: Homeowner Associations and the Rise of Residential Private Government.* New Haven, CT: Yale University Press.

McKenzie, Evan. 2005. "Constructing the Pomerium in Las Vegas: A Case Study of Emerging Trends in American Gated Communities." *Housing Studies* 20: 187–203.

Miller, Gary. 1992. *Managerial Dilemmas.* New York: Cambridge University Press.

Nelson, Robert H. 1999. "Privatizing the Neighborhood: A Proposal to Replace Zoning with Private Collective Property Rights to Existing Neighborhoods." *George Mason Law Review* 7: 827–880.

Nelson, Robert H. 2005. *Private Neighborhoods and the Transformation of Local Government.* Washington, DC: The Urban Insititute Press.

Olson, Mancure. 1965. *The Logic of Collective Action: Public Goods and the Theory of Groups.* Cambridge, MA: Harvard University Press.

Ostrom, Elinor. 1998. "A Behavioral Approach to the Rational Choice Theory of Collective Action." *The American Political Science Review* 92: 1–22.

Perry, Elizabeth J. 2008. "Chinese Conceptions of 'Rights': From Mencius to Mao – And Now." *Perspectives on Politics* 6: 37–50.

Polishchuk, Leonid, and Ekaterina Borisova. 2010. "Performance Assessment of Russian Homeowners Associations: The Importance of Being Social." Munich Personal RePEc Archive, July 2010.

Read, Benjamin L. 2003. "Democratizing the Neighbourhood? New Private Housing and Homeowner Self-organization in Urban China." *The China Journal* 49: 31–59.

Read, Benjamin L. 2008. "Assessing Variation in Civil Society Organizations: China's Homeowner Associations in Comparative Perspective." *Comparative Political Studies* 41: 1240–1265.

Rosenblum, Nancy L. 1998. *Membership and Morals: The Personal Uses of Pluralism in America.* Princeton, NJ: Princeton University Press.

Southern Metropolis Daily. 2013. November 6. Accessed October 25, 2014. https://www.chinadia logue.net/authors/304-Southern-Metropolis-Daily

Sun, Xiaoyi, and Ngai-ming Yip. 2014. "Managing the Nouveaux Riches: Neighborhood Governance in Upmarket Residential Developments in Shanghai." In *Housing Inequality in Chinese Cities*, edited by Y. Q. Huang and S. M. Li, 217–233. Abingdon: Routledge.

Tomba, Luigi. 2005. "Residential Space and Collective Interest Formation in Beijing's Housing Disputes." *The China Quarterly* 184: 934–951.

Webster, Chris, Fulong Wu, and Yanjing Zhao. 2006. "China's Modern Gated Cities." In *Private Cities: Global and Local Perspectives*, edited by G. Glasze, C. Webster, and K. Frantz, 153–169. London: Routledge.

Wu, Fulong. 2002. "China's Changing Urban Governance in the Transition towards a More Market-oriented Economy." *Urban Studies* 39. 1071–1093.

Wu, Fulong. 2006. "Transplanting Cityscapes: Townhouse and Gated Community in Globalization and Housing Commodification." In *Globalization and the Chinese City*, edited by F. L. Wu, 190–207. Oxford: Routledge.

Yip, Ngai-ming. 2012. "Walled without Gates: Gated Communities in Shanghai." *Urban Geography* 33: 221–236.

Yip, Ngai-ming, and Ray Forrest. 2002. "Property Owning Democracies? Home Owner Corporations in Hong Kong." *Housing Studies* 17: 703–720.

Yip, Ngai-ming, and Yihong Jiang. 2011. "Homeowners United: The Attempt to Create Lateral Networks of Homeowners' Associations in Urban China." *Journal of Contemporary China* 20: 735–750.

Zérah, Marie-Hélène. 2007. "Middle Class Neighbourhood Associations as Political Players in Mumbai." *Economic and Political Weekly* 42: 61–68.

Creating and defending concepts of home in suburban Guangzhou

Dan Feng[a], Werner Breitung[b] and Hong Zhu[a]

[a]Cultural Industry and Cultural Geography Research Center & School of Tourism Management, South China Normal University, Guangzhou, China; [b]Department of Urban Planning and Design, Xi'an Jiaotong-Liverpool University, Suzhou, Jiangsu, China

This paper uses the concept of "home" to analyze recent urbanization processes in suburban China from a cultural geography perspective. Urban growth, land development, and human mobility have had great impacts not only on land use and the built environment, but also on people's concepts of identity and belonging. As the current Chinese leaders have turned to explicitly viewing urbanization also as a social project, we need to understand better how it works as a social and cultural process. While Chinese societies have always conceived of "home" as something very stable, we now see this concept challenged in two ways: by human mobility and by the transformation of the places themselves. Suburban Guangzhou is a case in point. There are three population groups – the residents of gated communities, villagers, and rural migrants – that are all displaced in different ways. They settle side by side at the fringes of a city which most of them do not fully consider as their home yet. Interesting in this situation are their competing and interrelated claims of home. Being local or migrant, rich or poor, and urban or rural people, each of these groups has different resources and different strategies to construct, reconstruct, and defend their sense of home in this transient and contested space. Based on many years of research in Guangzhou, the authors try to disentangle these strategies and resource configurations. One aspect receiving particular attention is the role of spatial boundaries and of processes of exclusion and inclusion in this context.

1. Introduction

China is a country under enormous social and spatial change. As the map of opportunities – economic, social, cultural – is changing, people have become mobile in many ways. They move to the booming mega cities, adopt multi-local lifestyles, and become also socially more mobile. At the same time, places everywhere are also undergoing fundamental change. These dynamics represent chances, but also challenges, for people, businesses, and localities, and to the society as a whole. One area with huge and often underestimated implications is the need of many people to readjust their concepts of home and belonging.

In the Chinese society, home has always been an extraordinarily important and stable concept. Most people can trace their "home" to a place of ancestral roots, which plays a vital role for their identity, as a self-concept, and as they are viewed by others.

Even if individual people have moved, the concept of home as a place and in terms of social belonging has largely persisted.

This is changing now because of the sheer amount of people moving away from their ancestral home, because of their long-term perspective as members of new urban societies, and also because of the unprecedented degree to which the places themselves are changing under the forces of modernization and urbanization. With our research in suburban Guangzhou, we want to shed light on the multitude of ways that "home" is reconstructed in this setting and on the interrelatedness of these ways. The setting we chose is intriguing because of the overlap of competing and interrelated claims of home and the mechanisms how these claims are negotiated and enforced.

This setting is characterized by the multiple juxtapositions of commercially built gated communities and the village communities that used to own the land taken up now by these estates. The population of this area broadly consists of three groups – first, the residents of the gated communities, most of them middle-class migrants from other cities of China and beyond; second, the original villagers, who are actually the most local and immobile group, but who see their ideas of home vanishing and transforming as their villages are being urbanized; and third, a considerable population of rural migrants, who rent spaces in the villages and many of whom take up jobs serving the gated communities.

For all three groups, home is a vital issue. They are all displaced in different ways, more or less actively driven by their aspirations of social ascent, and now settling side by side at the fringes of Guangzhou. Neither the place nor the people have Guangzhou identities yet, but they owe their new opportunities to the growth of Guangzhou. While the forces of urbanization are physically and administratively merging the originally rural place into the mega city, new identities and new senses of place are emerging. By studying how people become urbanites and how they construct new concepts of home in this transient and contested space, we want to better understand urbanization as a social and cultural process. One aspect we pay particular attention to is the role of spatial boundaries and of processes of exclusion and inclusion in this context.

As the Chinese Government turned to actively promoting the urbanization of places and the creation of a larger urban middle-class, they based their policies on assumptions about the relationship of people and places, as expressed in concepts like ownership, citizenship, and belonging. It is therefore very timely to get a better understanding of these relationships and their roles in the ongoing urban and social transformations.

2. Geographies of homemaking and border-drawing

Home is a rich and powerful term. It is not only a physical location as house, but is "invested with meanings, emotions, experiences and relationships that lie at the heart of human life" (Blunt and Varley 2004, 3). A wide range of literature has focused on the home. Influenced by the thinking of humanistic geographers, the description of home as a haven has been prevalent. Home has often been understood in association with ideas, such as privacy, intimacy, and domesticity (Rybczynski 1986), a place where human beings can retreat and relax (Moore 1984). However, this positive and affective understanding of home has been subject to critique, in that the experience of home is not homogeneous. It is an ideal place for some, but it can also be a site of struggle and conflict for others (Brickell 2012). The home can be a space of belonging, intimacy, and desire, but it can also be a space of alienation, violence, and fear (Blunt and Varley 2004). Especially women and children are often subject to violence in home environments (Jones 1995; Brickell 2008, 2014).

Increasingly, the idea of the "ideal home" as a desirable, private, and comfortable space is being challenged as constructed by the dichotomy of inside and outside (Dovey 1985; Brickell 2012). Belonging is often achieved by excluding those who are viewed as aliens (Morley and Robins 1993; Schröder 2006, 30). This makes home a product of contestation for space. What is someone's home may at the same time become non-home to others. This logic is most obviously epitomized by the phenomenon of gated communities, which are especially prevalent in China. Claims of home in these communities are underpinned by the ownership of space and by the exercise of control over it.

Following the above understanding, home will be conceptualized in two ways. First, home is a combination of place, emotions, and sociospatial relations. Blunt and Dowling (2006, 2) point out that "home is a spatial imaginary: a set of intersecting and variable ideas and feelings, which are related to context, and which construct places, extend across spaces and scales, and connect places." Thus this article studies the home place and the feelings attached to home. These feelings and the related sociospatial relations are multi-scalar. They are not restricted to single housing units, but can refer to the neighborhood, home town, or the whole nation (Duncan and Lambert 2004, 388; Blunt and Dowling 2006, 29). In fact, home, understood as an imagination, as an idea, does not need to refer to a concrete geographical scale. What interests us in this paper is how homemaking processes transcend the domestic and occur in relation to outside systems, institutions, and communities.

Second, home then is also political, as it is a platform for the exercise of power, hegemony, and exclusion. Sibley (1995, 91), in this context, points to deep fear of difference. The perception of non-conforming people, activities, and artifacts may trigger acts of spatial purification and the reinforcement of boundaries. Such difference may be based on a variety of categorizations, such as gender, race, ethnicity, or class (Blunt and Varley 2004; Mallett 2004). In China, key differences in the society are wealth vs. poverty, rural vs. urban, and migrant vs. local. These differences are institutionalized in different modes of neighborhood governance and increasingly manifest in the bounded nature of residential space (Breitung 2012).

This article focuses on the material boundaries, such as gates and walls built around the gated communities, and explores the ways in which the boundaries are associated with power and identity. Such cultural analysis of the emergence of gated communities is rare. Most research focuses on the spatial, institutional, economic, and social dimensions of gated communities, such as explaining their proliferation, arguing about their social effects, and discussing their governance and planning (for a review, see Le Goix and Webster 2008). An important institutional argument, for example, is how homeownership is linked to a sense of belonging and governance participation through homeowners associations (e.g. Tomba 2005; Yip 2014; Breitung 2014).

As a cultural geographer, Pow (2007) argues that the new landscape of exclusion in China's gated communities reflects and represents the lifestyle and status of the Chinese middle-class in contrast to the world of uncivilized peasants and migrant workers outside the estate walls. From a consumerist perspective, gated communities have been described as Chinese dream homes (Pow and Kong 2007) and their promotion as a strategy of attracting the Chinese middle-class by promoting an idea of good life (Wu 2010). We will build on these cultural arguments in combination with a critical view on power relations when we investigate strategies of making and defending home in a society increasingly split into rich vs. poor, urban vs. rural, and local vs. migrant. The aim is to interpret home as a spatialized and politicized concept within China's specific social context.

To achieve this, it would not be enough to investigate gated communities and their populations in isolation. We must view them in context, as identities and spatialized concepts of home are formed in contrast and in competition to those of other groups living outside the gates. So, our empirical study in Guangzhou has from the outset included adjacent neighborhoods, which normally receive only inadequate attention.

3. Case studies and research methods

To understand the homemaking process in suburban China, this article draws on a decade of research in Guangzhou, the South Chinese metropolis and provincial capital. Guangzhou witnessed an early and rapid development of gated commodity estates. Especially the large-scale suburban mega estates have become models for other cities as their developers expanded their business throughout the country. Guangzhou and the surrounding Pearl River Delta also became a well-known example to study the politics of land conversion and in situ suburbanization (Lin 2009). From the population side, Guangzhou became one of the most prominent migrant cities in China, with domestic migrants (without Guangzhou *hukou*) accounting for 37.5 percent (National Bureau of Statistics 2010), plus numerous citizens with Guangzhou *hukou* but different origin. In the discourse reflected in the media, the people living in the huge suburban commodity housing enclaves are viewed as "new urban migrants." In contrast to migrant workers, these urban migrants normally have white-collar jobs and an urban *hukou*.

Within the broader Guangzhou area, we chose the most prominent case for suburbanization: Panyu, a former rural county turned district of Guangzhou. Our two case studies there are Nanpu Island and Shunde Country Garden (SCG), the latter actually developed from neighboring Shunde but stretching into Panyu. On Nanpu Island, we collected data in Lijiang Garden (LG), a commodity housing estate, and in the adjacent Jushu Village. SCG was investigated in conjunction with the neighboring Sangui village. The two cases were also chosen to represent suburban residential developments of different socioeconomic composition.

Nanpu Island, near the urban fringe of Guangzhou, is surrounded by three branches of the Pearl River. During the past decades, four big estates have been developed on this island. These estates are located on the eastern, better-connected part of the island, and most villages are in the west (Figure 1). Only Jushu village is caught in the middle of two gated housing estates. LG is the earliest and most-established project on this island. It is a large estate of about 10,000 households. As a whole, it is not strictly gated, but subdivided into subunits with tighter access control. LG is dominated by high-rise buildings. Only a few more exclusive subunits feature low-rises. Our survey in LG confirmed that many residents there are middle-class urban migrants with an urban household registration (*hukou*) from places other than Guangzhou.

The villages on Nanpu Island have existed for generations, with villagers engaged in farming and fishing before the gated communities were built on their agricultural land. While most of the farmland was expropriated (with compensation) by the Panyu government and sold to developers, the village land is still under collective ownership of the village community. Many villagers have replaced their old rural houses by more spacious ones and derive their main income from renting out space to migrant workers. The designs of the new houses are markedly different from the old ones, incorporating "urban" style elements, occasionally clearly borrowed from the architecture in the adjacent gated communities. Some of the new village residents who have moved into the rented spaces have started businesses or found work in the surrounding estates.

Figure 1.   Walled landscape on Nanpu Island.
Source: Author's drawing.

Despite such linkages, Nanpu Island has, in this process, changed into a fragmented place, composed of very different lifeworlds, juxtaposed to each other and separated by all kinds of walls and gates (Figure 1). The walls range from low-perimeter walls (<2 m) to high, solid, and continuous security walls. On some walls, sharp-pointed objects are added to prevent people from climbing over them. From the authors' observation over the past 6 years, the impermeability and access control have increased with more walls built or heightened. The gates installed by the developers of the commodity housing estates vary among symbolic ones, guarded control gates, and electronically operated gates (for typologies, see Luymes 1997). Most of the gates use electronic operation and guards at the same time. Sometimes, the entrances for cars and those for pedestrians are separate.

SCG lies a bit farther to the southwest at the administrative border of Guangzhou. It is a very large estate (369 km^2) that houses about 16,000 households with predominantly American-style low-rise buildings. Many are second homes or retirement homes of Hong Kong residents. This is typical of earlier developments in suburban Guangzhou. In our survey, 41 percent of the respondents were non-mainland China citizens, and the proportions of elderly residents and of employees with higher-ranking job positions were relatively high. SCG is an upper-middle-class community. Similar to the Nanpu Island case, it is surrounded by pre-existing villages. Sangui is the biggest of them.

This article draws on both surveys and in-depth interviews conducted between 2007 and 2010. In the two selected estates, 21 semi-structured in-depth interviews and a quantitative survey based on 538 questionnaires (263 in LG, 275 in SCG) were completed. In addition, participant observation and about 40 qualitative interviews were conducted in areas surrounding these estates. The respondents for the quantitative survey were selected at public places, which probably resulted in a not totally representative sample, although steps were taken to ensure a balance regarding age, gender, and activities performed. The profile of the survey samples can be seen from Table 1.

The subjects for qualitative interviews were selected from the respondents who agreed to leave their contact information during the survey. The in-depth interviews were semi-structured and guided by a list of questions. Their duration varied from 30 to 90 min. The interview quotes in this text are direct translations from Chinese. In addition, the lead author resided in Jushu village on Nanpu Island and conducted ethnographic fieldwork over one month, which enabled us to observe the daily life of the villagers and migrants in Jushu village. News reports and the Internet forums of the gated community residents were additional data sources.

Table 1. Profile of the survey sample in the two estates.

		LG (percentage)	SCG (percentage)
Gender	Female	55	49
	Male	45	51
Age	16–20	4	8
	21–30	39	25
	31–40	30	14
	41–50	7	11
	51–60	7	19
	61+	13	23
Education	Primary or junior secondary school	3	5
	Vocational or senior secondary school	16	25
	Non-academic further education	23	15
	Bachelor's degree	48	40
	Master's degree or higher	11	15
Work situation	Study	2	7
	Work	70	44
	Housewife, unemployed, long-term leave	13	12
	Retired	15	36
Hukou	Guangzhou urban	46	20
	Non-Guangzhou urban	46	31
	Guangzhou rural	2	1
	Non- Guangzhou rural	3	8
	Non-mainland China citizens	4	41
Length of living in Guangzhou	Grew up there	15	
	More than 10 years	24	
	3–10 years	40	
	Less than 3 years	21	

4. Home, power, and identity in suburban Guangzhou

4.1. Promoting the idea of home

In this section, we explore the strategies of how the imagination of the ideal home is promoted by developers and managers of the gated communities, and we analyze how those strategies are related to the making of boundaries.

When the work units gradually receded from housing provision and commodity housing was successively established, it also fundamentally changed the meaning of housing for urban residents. The house is no longer only a necessity as shelter, but it embodies personal desires, aspirations, and social status (Pow 2007) as well as identity and belonging (Saunders 1990). These symbolic and cultural meanings are important for the emerging urban middle-class in their efforts towards class distinction. The idea of home is also important in the context of residential mobility, as it creates or reassures the link between the self and the locale of one's residence. These two fundamental psychological functions should be kept in mind for the following analysis of individual or collective behavior in our case study setting. Before this, however, we need to widen our angle from the demand side to also incorporate the political economy of housing provision from the supply side. The symbolic and cultural meanings of home represent added value also in monetary terms. This added value is the foundation of the thriving real estate industry, on which much of the Chinese economy and local administration are based. So, we first need to understand the importance of "home" for industry and public policy.

Within their efforts of selling commodity housing and promoting homeownership, the concepts of house and home are often conflated (Mallett 2004). The developers market and mobilize the concept of home in order to promote their real estate. To that end, exaggerated projections of comfort, intimacy, and security are excessively transmitted through the images and texts presented in promotional brochures for gated communities. These do not merely attach emotional feelings of home to the individual unit; they always portray the bounded estate at large as home, with all the attributes described above attached to it. The boundary marks the spatial limits of the sense of home, and at the same time, it secures the imagination of a unified identity, constructed at the expense of those identified as "the other" (Young 1997).

The advertisements of LG, for example, portray the estate as a place of domesticity, comfort, intimacy, and trust. Figure 2 shows a happy couple with a baby enjoying their private time in the public space of LG. The rich fruits on the tree in front of them symbolize LG as a place of fruitful life – supposedly including family life and economic success. The most important elements of home, comfort and intimacy (Rybczynski 1986), are vividly conveyed by this figure. The text implies that good social relations and a happy family are bound up with the place. Figure 3 romanticizes the place with a poetical story and image. It promises an alluring combination of social relations and privacy, both of which are critical issues for people moving to commodity housing estates (Zhu, Breitung, and Li 2012). These two advertisements are very typical for the promotion of Chinese commodity housing. Most of the related promotion material is full of symbols of comfort and intimacy as well as sophistication and success.

The institutions managing LG, e.g. property management company and residents committee, are actively engaging in homemaking practices. The annual cultural and art festival, held by the management company aims at promoting the idea of LG as "home" to its inhabitants, as the rhetoric used in the promotion material for the tenth of these festivals illustrates:

Figure 2. Advertisement of LG. The text translates as "Here, someone harvests the fruit of love, someone makes friends by playing cards, and more find infinite joy."
Source: www.wzsky.net/html/60/33671.html
The present author has made reasonable efforts to trace the original owner of the advertisement. If you are or you know of the original owner of the advertisement, please contact the Journal to facilitate a correction to the record.

In Lijiang Garden, there is joy, there is happiness, there is quarrel and disappointment, but more of love and hope … because at this home, passion is thicker than water. As members of this home, we are responsible for its future … In the season when the flowers bloom, we are growing up with this home, we are thinking: how to make the flower of our lives bloom, and how to make the flower of this home bloom …

The developer links the concept of home with the place in both material and imaginary ways. However, what is marketed is not only the individual housing unit, nor is it the geographical place of Nanpu Island, but mainly the "brand" of LG – seemingly a newly created place. This is even more obvious at Country Garden, which for all its estates uses the slogan "Your Five-star Home," referring to both home and status. These pseudo-home places are produced as consumer-oriented, privatized, and controlled spaces, following the logic of Disneyworld (Relph 1991; Sorkin 1992; Warren 1996).

This goes so far that the existence of the outside space, especially the villages, is deliberately obscured and omitted from advertisements. In the publicized maps, the actual villages are often covered by fake esthetic images. Such construction of home also defines the identity of the residents without reference to village names. An identity of "LG people" (*Lijiang Ren*), for example, has been produced and articulated and widely appears in the media – instead of alternative descriptions, such as "Panyu people," "Nanpu people," or "Jushu people." The bounded residential space shapes a unified identity as a community from which the adjacent villages are excluded. Exclusion and

Figure 3. Advertisement of LG. The text translates as "Wait for you entering the elevator, help you with smile, no redundant words, we even forgot to ask each other's name."
Source: www.chinaadren.com/html/file/2006-6-22/2006622161543.html
The present author has made reasonable efforts to trace the original owner of the advertisement. If you are or you know of the original owner of the advertisement, please contact the Journal to facilitate a correction to the record.

inclusion are like two sides of one coin. Boundaries are drawn to foster a sense of community by defining who is an insider and who is an outsider. Spatial strategies such as boundary-drawing can influence social identities, which in turn also shape spaces (Sibley 1995). The ways in which the bounded spatial arrangements in our study area concretely reflect and produce social identities are elaborated upon in the next section.

4.2. Building a sense of home in the new estates

Although the making of "home" is embedded in social and economic contexts and influenced by external place-making and identity-creating efforts of developers, government propaganda, and neighborhood managers, it is essentially a process between the self and a place. "Home" and "sense of home" are subjective categories, and they can only be understood in detail on the subjective level. This section therefore examines the feelings and daily practice of the gated community residents.

4.2.1. Material and imaginary aspects of home

Residents from both LG and SCG stress their sense of home when talking about their connections to the commodity housing estates. Miss Li (LG resident), for example, described her feeling of home and her experiences related to this feeling:

> I have a strong feeling that this (Lijiang Garden) is my home, especially at the beginning when I just started to work. I was not quite adapted to the society then, and there were so many things to worry about. Every day I had to go to Guangzhou to work in the morning and back in the evening. I was wondering what the meaning of such an aimless life is. So I enjoyed coming back here. Every time when I cross the Lijiang Bridge and pass through the gate, the world quiets down at that moment like in a haven of peace (shiwai taoyuan). Even the air tastes sweeter.

In this sense, home is both material and imaginary (Easthope 2004; Blunt and Dowling 2006). The estate-as-home is a materially existing territory. Furthermore, material objects such as gates and the LG Bridge are of great importance as symbols of home. On the other hand, the estate-as-home is a set of feelings as well, when it is imagined as "haven of peace," where even "the air tastes sweeter." Besides this, "comfort," "safety," "privacy," and "peace" are the most frequent descriptors of emotion towards the estates of LG and SCG.

The scale of what people feel as home is not only their own flat, but the gated community at large, the estate-as-home, rather than apartment-as-home, in line with the message in the developers' promotion. Windy, a SCG resident, explained:

> There is a home living inside everyone's heart, and Country Garden is a large home for us … This is why we think Country Garden is successful. When you enter the gate of the estate, you are approaching a larger home, and when you open the door of your house you arrive at the smaller home.

The gates and walls thus geographically define the territory of home; meanwhile, they mentally symbolize the imaginary home, as in the above statement by Miss Li and in similar narratives by other residents, such as a SCG resident, who said *No matter which gate [of SCG] I am going through, I feel at home immediately.* As the gates, walls, and signs become the symbol of home, the boundary is not merely a material component of the gated community – more than that, it represents and shapes the imaginary of home. The removal of the broad range of gates and walls on Nanpu Island would therefore certainly cause strong objection.

On 26 August, 2010, an online protest against the removal of the neon light marking "LG" was launched by "the Lijiang people." Some residents noticed the change immediately and spread the news on the residential Internet forum of LG, which had around 75,000 members in 2010 and is popular with the outside public as well. The following posts demonstrate the close association of this material object with the LG inhabitants' sense and mental meanings of home.

> I went back to Lijiang Garden today. I took a glance: the word "Lijiang Garden" was gone. I felt lost. (NetID: jlxjd)

> Lijiang Garden is gone, with only an ugly iron frame left! Damn the urban management officers! (NetID: Herihuisui)

> The neon sign stands for Lijiang Garden. It used to be seen from both Xingguang Bridge and Panyu Bridge. It was my pilot lamp directing me home! (NetID: jxgzzl)

> It is not beautiful anymore. Previously I could see our home, "Lijiang Garden," from far away, but now … (NetID: Arlen)

4.2.2. The social and cultural construction of home in a new place

Home is constructed by Windy, the above-mentioned SCG resident, as a way to settle down:

> I am from Hong Kong, and my husband is from Britain. At the time of the Asian Financial Crisis, our company was affected, and our daughter was born in the same year. Therefore my husband considered changing his job to be a teacher. By chance I read a brochure called "Giving You a Five-star Home" (slogan of Shunde Country Garden). It touched my heart, because we were always longing to be settled, as we've been flying everywhere for our business. Neither Hong Kong nor the UK could be our home.

The majority of residents at SCG and (albeit to a lesser extent) at LG have migrated from other places to Guangzhou. This is important for understanding the significance of homemaking for the residents. Home is perhaps especially important for the Chinese, who tend to closely associate "home" with their roots (Su 2014). For those Chinese who migrate to other places, the making of new homes is particularly essential (Leung 2004). These should be places familiar to them, with communities to which they can belong. For the new arrivals to Guangzhou, it is easier to achieve familiarity and belonging in a newly formed community, rather than in an existing neighborhood or even in a village, in which they are viewed as outsiders; and it is easier when these communities are clearly identified and bounded. Gates and walls can create such a place to serve as the new "shared collective imaginary home" (Wiles 2008) to these migrants – a comfort zone that belongs to them, not to the Guangzhou locals or the people of the Nanpu villages.

Home is not only a place to settle down; it is also a better place – invested with positive emotions. Home is socially constructed as a territory owned and used by a group of people with a similar background and lifestyle, constructed as though against the outside world. As Blunt and Dowling (2006, 22) elaborate: "... the material form of home is dependent on what home is imagined to be, and imaginaries of home are influenced by the physical forms of dwelling." In suburban Guangzhou, home is imagined as a refuge from a hostile outside world, as a quiet and relatively private space.

The existence of a material boundary reinforces the difference between inside and outside, and shapes what home should be and shouldn't be. As in Miss Li's statement, LG is an ideal home place because it is different from the complicated society and frustrating workplace. It is not only that many "insiders" are migrants without local roots; also, "the world outside" is "complicated," meaning diverse, in terms of the three dimensions mentioned (urban–rural, rich–poor, and local–migrant). This understanding of home as a refuge for people to retreat from the outside world is founded on the distinction between private and public, home and work, and inside and outside world (Wardhaugh 1999). In a suburban setting, the difference between inside and outside is amplified by another very significant distinction: between the rural and the urban.

4.2.3. The construction of urban inside and rural outside

The firmly institutionalized rural–urban divide in China plays a significant role in the imaginaries of suburban space (Pow 2007). There exists a discursive boundary between rural and urban: rural places are generally considered inferior to cities. People living in rural places are viewed by urbanites as poor, unenlightened, and uncivilized. However, estates such as LG and SCG, although located in suburban Guangzhou, are still imagined as "urban." The desire to separate ideal homes from the inferior surroundings is quite strong, as a SCG woman emphasized:

> I think it is good (to have the walls and gates); otherwise Shunde Country Garden will get spatially integrated with the surrounding villages. You do not feel that you are living in a village here, although Shunde Country Garden is far from the city.

The rural environment is viewed not only as outside the scope of home, but also as the counter-imaginary by which the own urbanity and imagined superiority are constructed. A 70-year-old SCG resident compared the estate with the adjacent villages:

> The environment in Sangui, outside the estate, is very bad. The dirt is stirred up by the wind. No one ever sweeps the dirty street, scattered with litter. The most unbearable thing is that people spit everywhere on the street. All this does not happen in the estate ... Vegetables and meat are a little bit more expensive inside the estate, but they are clean and sold at a marked price. The vegetables in Sangui Market are laid on the ground, and I don't dare to buy them ...

The differences mentioned here have clear emotional connotations of a desirable home place vs. a non-home place described as undesirable in terms of both the physical environment (dust) and social practices (spitting, display of merchandise). These differences reflect the image of the rural and urban divide in China (see Figures 4 and 5).

The estate-as-home is presented as an urban and civilized place – in sharp contrast to the surrounding non-home places. The difference is not only one of varying degrees of comfort and belonging. Many respondents go further and describe the outside as a space of fear.

> My husband likes to take a walk in the evening. Whenever he goes out, I keep telling him to stay within the estate and not to go out to Sangui, because my husband is a foreigner, and it will be out of my control if something happens to him outside. He also reminds me to stay inside Shunde Country Garden when I go out with our child to play, and to let him know when I go to Clifford Estate or Guangzhou. We all regard the estate as secure especially for people from other places like us. (Windy, Shunde Country Garden resident)

Safety is also viewed as the most important distinguishing factor between the inside and the outside by the estate residents (Table 2). In the deeper interviews, however, it became clear that "safety" here is not necessarily the absence of crime, but a subjective feeling that is more about the absence of social heterogeneity (Breitung 2012). The ideal home, in the estate residents' view, is not only a safe, civilized, and orderly place, but is also a socially homogeneous one.

The ideas of safety, order, conformity, and social homogeneity are constructed around the place of home by the creation and reinforcement of material boundaries. They regulate the physical access to the place. In most cases, the inhabitants of the surrounding villages are not admitted into the gated communities. More importantly, they are never admitted into the residents' concept of home.

4.2.4. Homemaking and identity formation

Home is defined as a place, but it also shapes and is shaped by the definition of who the members of home are. The material and imaginary home reflects wider social relationships, also symbolizing social status and differentiation (Marcuse 1997). In the words of Caldeira (2000, 263), "residence and social status are obviously associated, and the home is a means by which people publicly signify themselves." In the perception of the residents living inside gated communities, the gates and walls not only

Figure 4.   Streetscape in Jushu village.
Author's photo (2007).

separate space, they also serve to categorize social groups. Here again, the categories of rural and urban stand out. These categories seem to provide the residents with reassurance not only of belonging, but also of status.

   Though the gated community residents did not comment negatively about their peasant neighbors in the interviews, almost half of them think of themselves as being better in terms of income, education, social status, civilization (*suzhi*), and safety (Table 2). Rural people, in contrast, are particularly often portrayed as badly behaved and uncivilized.

I think the residents in the estate have a better sense of behaving and personal hygiene than the residents in Sangui village. (Shunde Country Garden resident)

Sangui is of rural identity (nongcun). The Hong Kong residents in the estate are well-behaved and have good manners. Even children do not run and jump up and down when climbing the stairs. (Shunde Country Garden resident)

Figure 5.   Streetscape in LG.
Author's photo (2007).

The Chinese concept of *suzhi* (civilization), encompassing ideas of education, good behavior, and sophistication, is viewed as more important for the distinction between rural and urban than the social status expressed in household income (Table 2).

Homemaking and the formation of associated identities are based on the assumption of social similarity among insiders and difference from outsiders. Inside the gates and walls, "The neighbors are good, and I feel comfortable getting along with them"

Table 2.   The estate residents' view of differences between inside and outside (percentages).

Responses	Sites	Residents inside	Neighborhood outside	Same/don't know
Safety	LG	63.0	3.1	34.0
	SCG	78.1	1.8	20.0
Civilization *(suzhi)*	LG	57.0	1.5	41.4
	SCG	55.6	4.4	40.0
Education level	LG	52.5	2.7	44.9
	SCG	50.0	4.0	49.1
Social status	LG	44.5	2.7	52.9
	SCG	42.1	2.9	54.9
Household income	LG	42.6	3.8	53.6
	SCG	42.3	3.3	54.4

Notes: The table refers to the question "Talking about residents living in this estate and residents living in the neighborhoods just outside, who of these are in a better position according to following factors?" All respondents were residents of the commodity housing estates Lijiang Garden (LG) or Shunde Country Garden (SCG).

(SCG resident), while the outsiders are dangerous. "Home" means the members belonging to it can trust each other and develop close relationships, and this is shaped in contrast to the group who does not belong to it:

> You never know what you might encounter outside the estate. There are different kinds of people on the public bus. If I am on the shuttle bus of our community, I feel at ease because I think most of the passengers are from Shunde Country Garden.

Accordingly, contacts with the people in adjacent villages are avoided. According to our survey, most of the residents of the gated communities hardly know anyone in the adjacent neighborhoods (Table 3).

The differentiation of identity into inside and outside disconnects the neighbors living in the gated communities and outside, as evidenced by our interviews:

> I don't think there are intense relations between Shunde Country Garden and Sangui village. They are disconnected. The residents and their lifestyle are independent in Shunde Country Garden. Those living in the surrounding neighborhood are mostly peasants and migrant workers. It is a status problem. I think they have a different lifestyle.

> There is a gap between residents here and outside, and we can hardly have a common topic. People outside are afraid of talking to me. For example, when I want to buy a duck from them, they sell it to me, but we just smile, what can we talk about? We are on different levels. The education background and the style of conversation are quite different. I have a feeling that the person may intend to cheat me when he is too friendly to me. It does not matter if I am friendly to him.

However, difference and lack of contacts are not viewed as problematic. There are few conflicts, and many residents described the relationship as "fairly good." The dominant feeling, however, is indifference (Table 4).

The above analysis sheds light on the connection between home, boundary, and identity. The identity of the gated community residents living in suburban areas is based on the differentiation of inside/outside, at-home/not-at-home. The place inside the walls is imagined as an ideal home where its "family members" feel safe and comfortable, while the outside is an unhomely place. The people living inside the walls share a collective identity as family members and different from the outsiders. The gates and walls are spatially and socially produced to secure and strengthen the distinction. The differentiated landscape of home reflects wider social relations between local/migrant and urban/rural and also symbolizes the privileged middle-class status as well-educated and civilized.

Table 3. Acquaintance with people in adjacent neighborhoods (percentages).

Responses	LG ($n = 263$)	SCG ($n = 273$)
Hardly know	70.0	62.6
Not acquainted	16.7	22.3
Fairly acquainted	11.4	11.4
Very well	1.9	3.7

Notes: The table refers to the question "How acquainted are you with your neighbors living nearby LG/SCG?" All respondents were residents of Lijiang Garden or Shunde Country Garden.

Table 4. Relationship with people in adjacent neighborhoods (percentages).

Responses	LG ($n = 263$)	SCG ($n = 273$)
Tense	0.8	0.8
Some conflicts	0.8	0.4
Indifferent	63.5	55.5
Fairly good	30.0	38.3
Very good	4.9	5.1

Notes: The table refers to the question "How would you describe your relationship with the residents living nearby LG/SCG?" All respondents were residents of Lijiang Garden or Shunde Country Garden.

4.3. Loss of home – local villagers outside the estate

In-depth interviews with 29 residents living on Nanpu Island outside the estates reveal that none of them feel "unwanted" or discriminated against, which is unlike what has been reported, for example, from South Africa (Lemanski 2006). They generally accept or even approve of the boundaries.

> They set walls on their land. It's none of our business. I don't think the walls and guards are against us villagers. The arrow is toward the people intending to get in to collect garbage. Additionally, there were greedy and barbaric people. So, the guards have the duty to stop those shabby elements to get into the residential areas. We villagers are permitted to get in for a stroll if we behave well. (Jushu villager)

When the interviewees talk about security issues, they do not mean the local villagers. So, there seem to be no hard feelings attached to the boundaries that spatially separate their villages and the gated communities. The villagers rather consider the improvement of the infrastructure and services on Nanpu Island as positive consequences of the arrival of gated communities. The villages do share some of the benefits of urbanization. In particular, the landscapes created by the estates, such as lawns, fountains, and bronze statues, meet the villagers' imagination of modernity. In a family album, we found the backdrops of the photos to be mostly either their own home or the estate park (see Figure 6(a) and (b)). This shows some pride in the estate, in being part of the modern world that has intruded on their village and was actually constructed on their former land. Even though they are not directly included, some, as the quotation above illustrates, emphasize that with a good relationship to the guards, villagers could enter the estate and are treated differently from other outsiders. In addition, the villagers economically benefit from the gated communities adjacent to them: (1) they received a share of the compensation money after losing the land; and (2) they earn money from renting rooms to migrants who work as personnel of the estates or provide service to the residents of estates.

The overall attitude among the villagers towards the estates is consequently not too bad. However, this changes when their feelings of home are concerned. The estates have significantly reduced the area viewed as home by the villagers. They do consider their village, including the houses, farm land, and surrounding physical settings as their home place. Both LG and SCG have privatized the waterfront, fenced it, and blocked access to it for the villagers. The villagers can no longer enjoy the high-quality landscapes, which they experience as a deep loss:

Figure 6. (a, b) Photos taken in front of Country Garden on Nanpu Island.
Source: Copied from a family album in Jushu Village.

> It was comfortable to walk along the river. The river is now inside Shunde Country Garden, and we don't have a place for a walk anymore. There is little public space for recreation now. There used to be a large area of bamboo groves along the river, and the bamboo groves once were well-known. But everything disappeared after Shunde Country Garden came. (Sangui villager)

The loss of space and landscape triggers complaints also directed at the estates:

> The landscape was better in the past (before the arrival of Shunde Country Garden). There was river and sand, and the landscape was good. Now Shunde Country Garden blocks the view, and Sangui village is isolated. The famous eight beautiful landscapes of Sangui are destroyed. (another Sangui villager)

He feels it is unfair to stop them from getting inside of SCG because "it used to be our land and we should have that right." In reality, it is the estates that have the power of "mapping" the place, and they do generally exclude the villagers. This exclusion and the related conflicts extend to the services attached to membership. A Jushu villager argued that the space where the estates locate was theirs, so "Why don't they let us (Jushu villagers) take the shuttle bus (run by the estate)?"

Another change comes with the loss of the rural life. The transformation of the land relieves the adjacent villagers from their backbreaking labor on the farms and brings new economic opportunities, but at the same time it challenges the community life. The bonds between the villagers are perceived to be damaged. A Jushu villager, for example, expressed nostalgia for the countryside life:

> Under the big banyan tree at the head of the bridge was a gathering place for the villagers previously. It used to be very bustling, especially at night. We enjoyed the cool air of summer and chatted together. We had many conversation topics, such as the farm work, the dragon boat competition, and so on. At that time, we moved our dinner table to the outside and put it together with the other families. Nowadays, everyone is busy with his own business.

A Sangui villager confirms this:

> Everyone nowadays is busy making money, so the interaction between families is less than before. In the past, we always got together ... When we used to work in the collective production team, we had the same, and we spent much more time chatting after work.

He cherished the memory of the old days, and was frustrated by the loss of intimacy between the families in the village. The physical environment of the home place is invaded by the estates – so are the feelings of the village as a big home.

Finally, the divisions between rich and poor, rural and urban as well as local and migrant are perceived by the villagers as well. The luxury gated enclaves adjacent to the villages make the villagers differentiate themselves from their new neighbors, as Mr. Lu (Jushu villager) emphasized:

> I think they must be rich. Otherwise how do they have the large amount of money to buy such luxurious flats? Only the property management cost is around 300 RMB per month, which is our living expense for more than ten days.

He stressed that his business of selling flowers and plants on Jushu market targets the estate residents.

> What the villagers are concerned with is how to make money to support their family. They seldom buy flowers and plants except for orange trees as decoration before Chinese New Year. The residents of the nearby estates already have enough money. So, they grow plants for entertainment and to fill their boring life time.

Beyond the differentiation by economic status, the image of "uncivilized peasants" is also reinforced by the co-existence and spatial separation. Some villagers used the terms "rude" and "uncivilized" to describe their own people. Mrs. Guo, running a clinic on the market, said that "Only a few residents of Country Garden come to my clinic. How can the ladies who always spend hundreds of Yuan to make up their hair like my small clinic? They prefer big hospitals".

To some extent, the gap between rural and urban residents could be viewed as class differences, but it is more complex than this. It is more about *suzhi* and perceived sophistication than about financial wealth. The original villagers are not poor. They benefit a lot from their right to the land and the ability to rent out places to migrant workers. In fact, many of them do not need to work anymore but rely on a steady rental income.

The local vs. migrant division is therefore as important as the rural vs. urban and the rich vs. poor division. Home ownership in the village is based on local roots, not on wealth. In China, property rights for urban commodity housing can only be leased for a maximum of 70 years, but villagers, as a collective, actually own their land and the house built on it. This is the most important privilege local villagers have. Some of them emphasized that for this reason they would not purchase property in a fancy-looking estate even if they have enough money.

For them, home ownership is a major resource because it provides for a steady source of income. Beyond this, being local is also a social and cultural resource. The village-based social networks and their sense of place and belonging as locals can provide opportunities and stability during the dramatic suburban transition (Mallett 2004).

4.4. Migrant workers – living at the margins

Besides the estate residents living inside and the local peasants outside the walls, there exists a large group of people who frequently cross the boundaries – the migrant workers. This term refers to people who have moved away from their homes without being able to transfer their official residence registration (*hukou*) to their new place of residence (Gaetano and Jacka 2004, 1). Most of them are of countryside origin. They work for the lowest wages, not only in the manufacturing and construction industries, but increasingly in the service sector, often in occupations that urbanites shun. In our cases, migrant workers are attracted by job opportunities created by the gated communities, as service workers for the estates or growing and selling vegetables, meat, or other goods to the estate residents. Some of them live in dormitories provided by the estates, but most migrants rent rooms in the villages. Almost every household in Jushu village rebuilt their house, adding more floors to accommodate migrant workers. According to the Jushu village committee, the village has only 920 local households, while there are approximately 2000 migrant workers as renters.

In China, migrant workers used to be called the "floating population" (*liudong renkou*), which suggests that they are on the move (Gaetano and Jacka 2004) and might eventually go back to their places of origin. Over time, however, it became evident that most of them preferred to find an urban place of residence to settle down. Therefore, homemaking has become a critical issue for this group too. The difference is that they often lack the resources to resolve this issue.

According to our interviews, their social networks initially influenced their decisions to take jobs at Nanpu Island and SCG, which conforms to other findings (compare Zhao 1994; Knight and Gunatilaka 2010). However, it is difficult for them to develop social relationships with either the estate residents or local villagers. A waitress working for a restaurant in Sangui village said that it is hard for them to get along with "local people" (Sangui villagers):

> They are rude. The Hong Kong residents (living in Shunde Country Garden) are nice. We have to pay much more attention when we are serving Sangui people. A tiny fault puts them off ... They will roar at us if they want more tea and sometimes they will break the cups. Hong Kong customers will not do this. They (Sangui villagers) look down upon us, and of course we will not treat them well. Finally they will go to complain to the boss. So we are really annoyed sometimes.

The villagers aim at securing their status, which is challenged from "above" (estate residents), by confirming hierarchies and establishing borders to those with lower status. Many local villagers expressed negative feelings towards the migrant workers directly. An old doctor (Jushu villager), who runs a clinic, stressed that she would do home visits for the estates residents, but not for the migrant workers, because she heard of cases of blackmail.

There is no clear physical boundary spatially separating local villagers from migrant workers in a way comparable to what gated communities do. A villager interviewed argued that "no one would want to rent our house if we treated him as a thief and asked him to present a card at any time ... The price of our house would no longer be higher than in neighboring villages". But in practice, the discursive boundary does exist. Even if the two groups live in the same building, they use separate entrances. The same is true for the use of public space. The park located in the middle of Jushu village is normally occupied by local people, while the migrant workers use another small park.

For some, these somewhat subtle bordering practices were not enough:

> If there were too many outsiders, all of the entrances and exits of our village could be used to check the permits. Different from the way that Country Garden (gated community on Nanpu Island) uses walls for protection, we could set security booths at each road leading to the village. Chigang village at Shilou town, where my older sister lives, has set such an example. There are four security booths at each of the four road entrance. The people need to show their ID card to get in. The migrants who rent places in the village are able to get this card from a local administrative department. In fact, a wall is not necessary for the village, because the thickly dotted houses have the same function as the walls around the estates. Our village wants to adopt this method all the time. Nowadays, there is a temporary security booth at the entrance of our village, but people don't need to show any card. Even if you are a stranger and behave well, you can get into our village. (Jushu villager)

The migrant workers are perceived as a "threatening difference" (Sibley 1995) by the local villagers. It is difficult for them to build social relationships between the groups. No matter how long migrant workers live in the village, their attachment to these places and feeling of home is weak. Some of them frequently cross the boundary between the estates and the outside to deliver goods, for example, but walking inside the estates makes them feel "out of place" (Cresswell 1996). These migrants are indeed "floating" between places where they do not feel at home.

5. Discussion

The above investigations have shown a very diverse set of social groups with different resources and different backgrounds who consequently pursue different strategies of homemaking. Figure 7 illustrates their characteristics according to the three dimensions urban vs. rural, rich vs. poor, and local vs. migrant. The migrant workers (renters in the village) differ from both other groups (estate residents and owners in the village) by being poor. This puts them in an inferior position to both, although most of the estate residents are also migrants, and the owners of the village houses share their rural background. The latter distinguish themselves as the only locally rooted group. This is an advantage they have even over the estates residents – but those can compensate for the lack of local roots by sufficient financial means and by their status as modern urbanites. Accordingly, the opportunities and strategies for homemaking differ among the three groups.

Residents of the gated communities follow the promises of the real estate advertisements and invest money to avail themselves of the space and the nice physical and homogeneous social environment they want – so that they can nurture their dream home in it. Border-drawing is an essential strategy to defend the acquired space. Apart from territory, a second essential aspect of home for them is status. Moving home must come with social ascent; therefore, distinction from the rural surroundings is crucial.

The villagers already own the space they live in, and they build on existing cultural and community links. They have their own resources and were able to grab opportunities that came with the real estate development to significantly progress economically. Yet their home space is severely transforming and they consequently struggle to maintain a sense of home. They gained from urbanization but suddenly find themselves in a position of inferiority relative to the gated community residents. Their claim of ownership and belonging to the place is challenged by the walls of the estate and simultaneously by their lesser ability to demarcate their space against the incoming migrant workers.

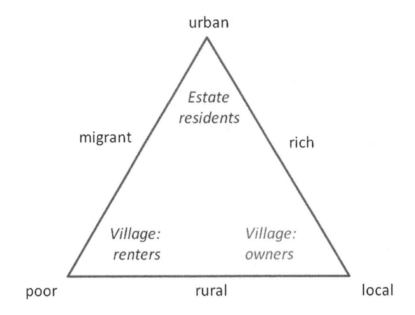

Figure 7.   Differences between population groups in the study area.
Source: Author's drawing.

The latter face the greatest difficulties because they lack all resources – monetary, cultural, and institutional – and they shuttle between the worlds without feeling at home in either of them. In both of these worlds, they are confronted with exclusionary practices of more powerful groups, who themselves struggle with their own sense of home. In many cases, these migrant workers will not be able to establish a new long-term home in their current place, but might eventually move on to urban Guangzhou or other places.

Space and the reassurance of social identity are key factors of the observed home-making strategies. Space is acquired, demarcated, purified, and symbolically charged. Ownership has become a powerful tool for homemaking in contemporary China. Ownership is seen as a legitimate justification for practices of fencing and exclusion. Ownership can also be a lever for attaining the status as urban resident (*hukou*). The construction of social identities is the second key strategy in homemaking and becoming an urbanite. Symbols of affluence, community, and lifestyle play an important role in linking place and identity. More importantly, practices of social purification and exclusion help to assure the connection between social identity and space. The legitimacy of these strategies is not really challenged, not even by those excluded. Rather, the villagers would like to adopt similar strategies. Although they can build their own claims on their traditional rights (especially landownership), they are clearly less powerful than those endowed with wealth and urban status.

Funding

This work was supported by the German Science Foundation (DFG) [grants number BR3546/1-2 and BR3546/2-1]; the National Science Foundation of China [grant number 41171123] (P.I.: Werner Breitung); as well as by the National Science Foundation of China [grant number 41301139] (P.I.: Dan Feng).

References

Blunt, Alison, and Robyn Dowling. 2006. *Home*. Abingdon: Routledge.

Blunt, Alison, and Ann Varley. 2004. "Introduction: Geographies of Home." *Cultural Geographies* 11: 3–6.

Breitung, Werner. 2012. "Enclave Urbanism in China: Attitudes towards Gated Communities in Guangzhou." *Urban Geography* 33: 278–294.

Breitung, Werner. 2014. "Differentiated Neighbourhood Governance in Transitional Urban China: Comparative Study of Two Housing Estates in Guangzhou." In *Neighbourhood Governance in Urban China*, edited by Yip Ngai-ming, 145–166. Cheltenham: Edward Elgar.

Brickell, Katherine. 2008. "'Fire in the House': Gendered Experiences of Drunkenness and Violence in Siem Reap, Cambodia." *Geoforum* 39: 1667–1675.

Brickell, Katherine. 2012. "'Mapping' and 'doing' Critical Geographies of Home." *Progress in Human Geography* 36: 225–244.

Brickell, Katherine. 2014. "'Plates in a Basket Will Rattle': Marital Dissolution and Home 'Unmaking' in Contemporary Cambodia." *Geoforum* 51: 262–272.

Caldeira, Teresa. 2000. *City of Walls: Crime, Segregation, and Citizenship in Sao Paulo*. Berkeley: University of California Press.

Cresswell, Tim. 1996. In *Place/Out of Place: Geography, Ideology, and Transgression*. Minneapolis: University of Minnesota Press.

Dovey, Kimberly. 1985. "Homes and Homelessness." In *Home Environments* (Human Behavior and Environment: Advances in Theory and Research 8), edited by I. Altman and C. Werner, 33–64. New York: Plenum Press.

Duncan, James S., and David Lambert. 2004. "Landscapes of Home." In *A Companion to Cultural Geography*, edited by James S. Duncan, Nuala C. Johnson, and Richard H. Schein, 382–403. Malden, MA: Blackwell.

Easthope, Hazel. 2004. "A Place Called Home." *Housing, Theory and Society* 21: 128–138.

Gaetano, Arianne, and Tamara Jacka. 2004. *On the Move: Women and Rural-to-Urban Migration in Contemporary China*. New York: Columbia University Press.

Jones, Gill. 1995. *Leaving Home*. Buckingham: Open University Press.

Knight, John, and Runatilaka Gunatilaka. 2010. "Great Expectations? The Subjective Well-being of Rural–Urban Migrants in China." *World Development* 38: 113–124.

Le Goix, Renaud, and Chris Webster. 2008. "Gated Communities." *Geography Compass* 2: 1189–1214.

Lemanski, Charlotte. 2006. "Spaces of Exclusivity or Connection? Linkages Between a Gated Community and Its Poorer Neighbour in a Cape Town Master Plan Development." *Urban Studies* 43: 397–420.

Leung, Maggi W. H. 2004. *Chinese Migration in Germany – Making Home in Transnational Space*. Frankfurt: IKO.

Lin, George. 2009. *Developing China: Land, Politics and Social Conditions*. London: Routledge.

Luymes, Don. 1997. "The Fortification of Suburbia: Investigating the Rise of Enclave Communities." *Landscape and Urban Planning* 39 (2–3): 187–203.

Mallett, Shelley. 2004. "Understanding Home: A Critical Review of the Literature." *The Sociological Review* 52: 62–89.

Marcuse, Peter. 1997. "The Ghetto of Exclusion and the Fortified Enclave: New Patterns in the United States." *American Behavioral Scientist* 41: 311–326.

Moore, Barrington. 1984. *Privacy: Studies in Social and Cultural History*. New York: Sharpe.

Morley, Dave, and Kevin Robins. 1993. "No Place Like Heimat: Images of Home (Land) in European Culture." In *Space and Place: Theories of Identity and Location*, edited by E. Carter, J. Donald, and J. Squires, 3–31. London: Lawrence and Wishart.

National Bureau of Statistics. 2010. "The 6th National Population Census." Accessed October 22, 2014. www.stats.gov.cn/ztjc/zdtjgz/zgrkpc/dlcrkpc/

Pow, Choon-Piew. 2007. "Securing the 'Civilised' Enclaves: Gated Communities and the Moral Geographies of Exclusion in (Post-)Socialist Shanghai." *Urban Studies* 44: 1539–1558.

Pow, Choon-Piew, and Lily Kong. 2007. "Marketing the Chinese Dream Home: Gated Communities and Representations of the Good Life in (Post-)Socialist Shanghai." *Urban Geography* 28: 129–159.

Relph, Edward. 1991. "Post-modern Geography." *Canadian Geographer* 35: 98–105.

Rybczynski, Witold. 1986. *Home: A Short History of an Idea*. New York: Penguin Books.

Saunders, Peter. 1990. *A Nation of Home Owners*. London: Routledge.

Schröder, Nicole. 2006. *Spaces and Places in Motion: Spatial Concepts in Contemporary American Literature*. Tübingen: Gunter Narr Verlag.

Sibley, David. 1995. *Geographies of Exclusion: Society and Difference in the West*. London: Routledge.

Sorkin, Michael. 1992. *Variations on a Theme Park: The New American City and the End of Public Space*. New York: Hill and Wang.

Su, Xiaobo. 2014. "Tourism, Modernity and the Consumption of Home in China." *Transactions of the Institute of British Geographers* 39: 50–61.

Tomba, Luigi. 2005. "Residential Space and Collective Interest Formation in Beijing's Housing Disputes." *The China Quarterly* 184: 934–951.

Wardhaugh, Julia. 1999. "The Unaccommodated Woman: Home, Homelessness, and Identity." *Sociological Review* 47: 91–109.

Warren, Stacy. 1996. "Popular Cultural Practices in the 'Postmodern City'." *Urban Geography* 17: 545–567.

Wiles, Janine. 2008. "Sense of Home in a Transnational Social Space: New Zealanders in London." *Global Networks* 8: 116–137.

Wu, Fulong. 2010. "Gated and Packaged Suburbia: Packaging and Branding Chinese Suburban Residential Development." *Cities* 27: 385–396.

Yip, Ngai-ming. 2014. "Neighbourhood Governance in Context." In *Neighbourhood Governance in Urban China*, edited by Yip Ngai-ming, 1–21. Cheltenham: Edward Elgar.

Young, Iris Marion. 1997. "House and Home: Feminist Variations on a Theme." In *Intersecting Voices: Dilemmas of Gender, Political Philosophy, and Policy*, edited by I. M. Young, 134–164. Princeton, NJ: Princeton University Press.

Zhao, Yaohui. 1994. "Leaving the Countryside: Rural-to-Urban Migration Decisions in China." *American Economic Review* 89: 281–286.

Zhu, Yushu, Werner Breitung, and Si-Ming Li. 2012. "The Changing Meaning of Neighbourhood Attachment in Chinese Commodity Housing Estates: Evidence from Guangzhou." *Urban Studies* 49: 2439–2457.

Index

Note: Page numbers in **bold** type refer to figures
Page numbers in *italic* type refer to tables
Page number followed by 'n' refer to notes

For Product Safety Concerns and Information please contact our EU
representative GPSR@taylorandfrancis.com Taylor & Francis Verlag GmbH,
Kaufingerstraße 24, 80331 München, Germany

Printed and bound by CPI Group (UK) Ltd, Croydon, CR0 4YY
01/05/2025
01858407-0002